Progress in Precision Agriculture

Series Editor

Margaret A. Oliver, Soil Research Centre, University of Reading, Berkshire, Berkshire, UK

This book series aims to provide a coherent framework to cover the multidisciplinary subject of Precision Agriculture (PA), including technological, agronomic, economic and sustainability issues of this subject. The target audience is varied and will be aimed at many groups working within PA including agricultural design engineers, agricultural economists, sensor specialists and agricultural statisticians. All volumes will be peer reviewed by an international advisory board.

More information about this series at http://www.springer.com/series/13782

Avital Bechar
Editor

Innovation in Agricultural Robotics for Precision Agriculture

A Roadmap for Integrating Robots in Precision Agriculture

 Springer

Editor
Avital Bechar
Institute of Agricultural Engineering
Agricultural Research Organization – Volcani
Institute
Rishon LeZion, Israel

ISSN 2511-2260 ISSN 2511-2279 (electronic)
Progress in Precision Agriculture
ISBN 978-3-030-77035-8 ISBN 978-3-030-77036-5 (eBook)
https://doi.org/10.1007/978-3-030-77036-5

This Springer imprint is published by the registered company Springer Nature Switzerland AG
The registered company address is: Gewerbestrasse 11, 6330 Cham, Switzerland

This book is dedicated to the memory of my father, Jacob Bechar.

An autodidact and a humble man who instilled in me the love of study and science.

Preface

This book is part of the new book series on Precision Agriculture established by Springer with Margaret Oliver as the series editor. There are very few books that deal with agricultural robotics and even fewer on robotics and precision agriculture. Until recently, research in the fields of agricultural robotics and precision agriculture evolved along parallel paths with little interaction or few relations between them. The seeds for this book were resulted from the BARD international workshop on "Innovation in Agricultural Robotics for Precision Agriculture" and the 10th European Conference on Precision Agriculture at the Volcani Center, Israel, in 2015. The aim of the book is to introduce agricultural robotics for precision agriculture, to present the conditions, rules and limitations, concepts, principles and required abilities for implementing robots in precision agriculture tasks and to review the state-of-the-art of agricultural robotics in different aspects of precision agriculture. The chapters were written by leading experts showing the links between agricultural robotics and precision agriculture. The book aims to guide readers in research, development and design of agricultural robots for precision agricultural tasks. All chapters include case studies to illustrate the techniques.

The book starts with an introduction in Chap. 1 on the agricultural sector, characteristics of the agricultural domain, agricultural robotics and the revolution that led to precision agriculture, and the Possibilities offered by robotics to precision agriculture. Chapter 2 overviews the principles, conditions and guidelines for agricultural robots to perform precision agriculture tasks, evaluate the requirements of robotic systems and presents associated concepts and the characteristics of the complexities and types of precision agricultural tasks from a robotic perspective. Chapter 3 provides an overview of the current sensors and data acquisition used by robots for precision agriculture and specifically modelling of crops, soils and other environments. There is also an appraisal of sensors and sensing principles, including several application case studies. Chapter 4 focuses on agricultural robots for precision agriculture tasks in orchards and analyses the incentives for developing robots for these tasks in that environment. It also reviews the various precision agriculture tasks in orchard and provides numerous case studies. Robotics for precision viticulture is described in Chap. 5 with the investigation of technological needs, barriers and current solutions for competitive vineyards; a review on the different precision agriculture tasks comprising precision viticulture and robotic solutions is provided. Chapter 6 provides a comprehensive discussion on the state-of-art of robotic spraying, with a review of weed and disease sensing tasks, and of precision actuation of treatments. The chapter shows how the building blocks of integrated robotic systems for precision crop protection are developing rapidly. Chapter 7 focuses on autonomous robot teams and addresses the integration of autonomous aerial inspection with autonomous ground intervention to perform precision agricultural tasks. The chapter describes the coordination, planning and monitoring mechanisms implemented, as well as additional interesting characteristics of the autonomous robot team. The book closes with emerging directions of precision agriculture and a summary of agricultural robotics. Chapter 8 provides a futuristic vision and aims to shed some light on the next steps in the evolution of precision agriculture and agricultural robotics and the technological factors that will drive this evolution.

Rishon LeZion, Israel   Avital Bechar
July 2020   Margaret Oliver, Book series editor

Contents

Contributors

Avital Bechar Institute of Agricultural Engineering, Volcani Institute, Rishon LeZion, Israel;
ARO, Rishon LeZion, Israel

Dionysis Bochtis Institute for Bio-Economy and Agri-Technology – IBO, Centre for Research & Technology – Hellas (CERTH), Thermi, Thessaloniki, Greece

Fernando Auat Cheein Department of Electronic Engineering, Universidad Técnica Federico Santa María, Valparaíso, Chile

Jesus Conesa-Muñoz Centre for Automation and Robotics, CSIC-UPM, Arganda del Rey, Madrid, Spain

Alexandre Escolà Research Group on AgroICT & Precision Agriculture, Department of Agricultural and Forest Engineering, Universitat de Lleida – Agrotecnio-CERCA Centre, Lleida, Catalonia, Spain

Manoj Karkee Washington State University, Prosser, WA, USA

Serafeim Moustakidis Institute for Bio-Economy and Agri-Technology – IBO, Centre for Research & Technology – Hellas (CERTH), Thermi, Thessaloniki, Greece; AIDEAS OÜ, Tallinn, Harju Maakond, Estonia

Ashwin S. Nair PRISM Center and School of IE, Purdue University, West Lafayette, USA

Shimon Y. Nof PRISM Center and School of IE, Purdue University, West Lafayette, USA

Roberto Oberti Department of Agricultural and Environmental Science – DiSAA, Università degli Studi di Milano, Milan, Italy

Angela Ribeiro Centre for Automation and Robotics, CSIC-UPM, Arganda del Rey, Madrid, Spain

Joan R. Rosell-Polo Research Group on AgroICT & Precision Agriculture, Department of Agricultural and Forest Engineering, Universitat de Lleida – Agrotecnio-CERCA Centre, Lleida, Catalonia, Spain

Francisco Rovira-Más Universitat Politècnica de València, Valencia, Spain

Verónica Saiz-Rubio Universitat Politècnica de València, Valencia, Spain

Ze'ev Schmilovitch Institute of Agricultural Engineering, Agricultural Research Organization, Rishon Lezion, Israel

Abhisesh Silwal Washington State University, Prosser, WA, USA

Qin Zhang Washington State University, Prosser, WA, USA

Chapter 1
Mobile Robots: Current Advances and Future Perspectives

Dionysis Bochtis and Serafeim Moustakidis

1.1 Introduction

The highly dynamic agricultural sector is driven by a range of factors while confronting new challenges, from food security to ecological concerns and land-use issues. First, the world's population is likely to increase from today's figure of 7.5 billion to an estimated 9.7 billion in 2050, with most of the current increase in developing countries and likely to remain so. Urbanization is another important factor that is expected to impose new patterns of food production, consumption and demand. It has also stimulated improvements in infrastructure, including cold chains that ensure and extend the shelf life of goods that are being traded nowadays. To provide high quality, affordable and safe foods for the growing world population poses a huge challenge given that the expected total demand for livestock products will almost double in sub-Saharan Africa and South Asia by 2050, whereas consumption in OECD countries will barely change as shown in Fig. 1.1.

Climate change is increasingly causing weather events such as higher temperatures, shifting seasons, more frequent and extreme climatic effects, flooding and drought, which also affect agricultural production (Lampridi et al. 2019a, b). Although climate change might also play a positive role in agricultural production locally, for example in the higher latitude regions, its overall effect on crop production is expected to be negative. The agricultural sector also faces a variety of additional challenges including:

D. Bochtis (✉) · S. Moustakidis
Institute for Bio-Economy and Agri-Technology – IBO, Centre for Research & Technology – Hellas (CERTH), 6th km Charilaou-Thermi Rd, GR 57001 Thermi, Thessaloniki, Greece
e-mail: d.bochtis@certh.gr

S. Moustakidis
AIDEAS OÜ, Narva Mnt 5, Tallinn, Harju Maakond, Estonia

© Springer Nature Switzerland AG 2021
A. Bechar (ed.), *Innovation in Agricultural Robotics for Precision Agriculture*,
Progress in Precision Agriculture,
https://doi.org/10.1007/978-3-030-77036-5_1

1

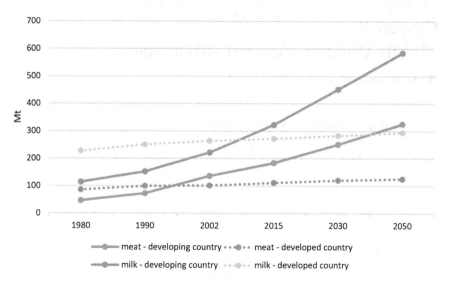

Fig. 1.1 Past and projected trends in total consumption of meat and milk in developing and developed countries. Data adapted from FAO: "Global agriculture towards 2050: High-level Expert Forum, Rome 12–13 October 2009"

(i)   the limited agricultural area available worldwide that can be increased only marginally,

(ii)   the rapidly evolving consumer habits in emerging countries,

(iii) the conflict between renewable energy and agriculture in terms of land use, and

(iv)   the limits on capacity of the agricultural machinery in highly developed countries.

Meeting the aforementioned challenges, a 2.4% rate of growth in crop production per year is required (that corresponds to a doubling of agricultural production by 2050 to feed the rapidly growing world population; this has been estimated in numerous studies (Godfray et al. 2010; OECD-FAO 2012; Tilman et al. 2011).

Automation technology generates considerable improvements in agricultural productivity, but at the same time it faces several technical issues as far as the management and use of resources in the production, the increased demands for product quality and organic farming products, the complexity of human–machine interactions, and the complementarity between labour and technology (Marinoudi et al. 2019). Information Communication Technologies (ICT) also play a key role in offering significant benefits in various areas such as field information through sensing, data analytics and autonomous robots to replace manual labour in agricultural activities (harvesting, weeding etc.). To reduce the gap between the installed and realised performance of agricultural machines, a holistic approach should be adopted taking into account the performance at both local (single machines) and global levels (the whole process). Single machine capabilities could be enhanced by integrating systems that provide assistance or autonomous robotic functions, or improvement in performance of the entire process could be accomplished by optimizing the interrelation between single machines.

1.2 Past, Current, and Future Phases in Agricultural Operations

Although the major developments during the first half of the 20th century that took place in industrial countries, the decisive break in the history of agricultural technology came after the 1950s in the second half of the century and at the beginning of the 21st century. The timeline of the evolutionary progress in agriculture is shown in Fig. 1.2.

Norman Borlaug's initiatives led to the Green Revolution, which in turn led to the renovation of agricultural practices that began in Mexico in the 1940s and spread worldwide in the 1950s to 1960s. It increased significantly the number of products and the number of calories produced per hectare of agricultural land. The green revolution was motivated by widespread concerns about hunger and rapidly growing populations in the world. The next big technological advancement was the introduction of rotary combines by Sperry–New Holland in 1975 allowing the crop to be cut and separated in one pass over the field. In the following years and especially in the 1980s on-board electronics were also integrated to measure threshing efficiency, enabling operators to obtain better grain yields by improving several critical operating properties (such as ground speed). The advent of Genetic Modification (GM) technology was another significant milestone in the mid-1990s that allowed the transfer of genes for specific traits between species using laboratory techniques that were adopted by growers of large acreage field crops worldwide. Satellite imaging was also introduced in the mid-1990s enabling farmers to monitor their land and increase crop yields through enhanced precision agriculture (Angelopoulou et al. 2019). It also facilitated improved access to land and provided information for better policy choices that benefitted both large scale and smallholder farmers. In 2000, the world's first touchscreen phone came out. Although they had fewer capabilities than the smartphones available today, they were very popular because they introduced a promising

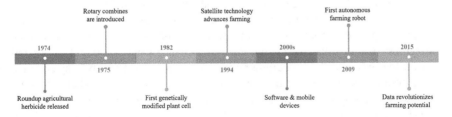

Fig. 1.2 Evolutionary change in agriculture: the past, present and future

technology for farmers so that they could stay connected to their colleagues, have access to data while on-the-go and be able to place orders for agricultural products (e.g. seed or fertilizer) at any place or at any time.

A step beyond mechanization is currently achieved through ICT and automation (Bochtis et al. 2014). The ICT-supported production systems decrease, to a large extent, the human requirements for sensor technology and reasoning about decisions to be made because they can increase the operational speed, the capacity of the system and the repeatability of various tasks and processes (Sørensen et al. 2010a, b; Sørensen et al. 2011). The ICT and automation technologies that have been successfully implemented in agricultural production include:

- Wireless technologies
- Global navigation systems
- Geographic information systems
- Management information systems
- Telematics systems
- Monitoring technologies, and
- Advanced control systems

Throughout all this period of transformation of the agricultural production system from the mechanization phase to the ICT phase, there was an increase in capacity with the greater effectiveness of the machines (Achillas et al. 2019; Seyyedhasani and Dvorak 2018; Sørensen and Bochtis 2010). On the other hand, this increase could be attributed to improvement in the efficiency of management tasks because they have been enhanced by the availability of information and the various decision support systems. The labour costs also decreased as a result of the automation of several work tasks. The new paradigm of the production system improved inputs to some extent, in terms of agrochemicals, with the introduction of precision agriculture principles (Tozer 2009). However, the production systems ended up by being more complex, requiring greater expertise of the farmers and also higher service costs. The level of investment of the agri-business also increased dramatically.

Big data and analytics are now revolutionizing many industries and are also turning the agricultural industry on its head. Nowadays, precision farming technology, empowered by big data and machine learning, allow farmers to crunch massive amounts of data collected by sensors to help them to make better-informed decisions in terms of resource usage, operational costs, optimal planning strategies and environmental impact. However, the introduction of big data into precision agriculture has brought a wide range of concerns among both farmers and agricultural data service providers about the privacy, ownership and use of farm data (Table 1.1). Farmers are mainly concerned about the uses to which their data might be put and how those uses could end up putting farmers at a competitive disadvantage relative to the companies with whom they are sharing their data. On the other side, data service providers have concerns about potential violations of intellectual property or licensing restrictions in relation to data that the farmers do not own. Data aggregation and related services for farming are expected to grow, therefore, it is clear that data

Table 1.1 Benefits and concerns of data revolution in agriculture

Benefits	Grey zone	Concerns
Regulatory compliance	Ownership	Transparent data protection plans
Improved monitoring	Liability & security	Ambiguous data-sharing regulations
Manageable operating costs	Collection, access & control	Compliance with government
Detailed yield tracking	Who profits of data?	Crop-field data affect crop prices
Reporting and work efficiency		Centralised data repositories
Sustainability reporting		
Mapping		

ownership and security will be among the main concerns. Despite all the aforementioned concerns and challenges, big data and analytics will continue to grow and advance, and to offer hope for feeding the world's population of tomorrow.

1.2.1 Robotics and Application Domains

1.2.1.1 General

Robots can be categorized on the following four key factors (Fig. 1.3) that also mark out potential robot markets: (i) operating environment (ii) interaction or collaboration with users, (iii) physical format and (iv) performing function.

Operating Environment: Five operating environments, namely: ground, air, water or underwater, space and inside the human body are the primary ones together with their sub-divisions (e.g. deep or shallow water, indoor or outdoor or underground). There are also types of robots that can operate in two or more different environments, for example on the surface of water with the ability to operate both underwater and in the air.

Interaction and Collaboration: By default, robots behave or perform tasks with a high degree of autonomy. Their tasks typically involve interaction with other users either remotely or with a specific communication link, whereas their level of autonomy for decisions can be categorised as (i) programmed, (ii) tele-operated or master slave, (iii) supervised, (iv) collaborative and (v) fully autonomous. Robots

Fig. 1.3 Categories of robots

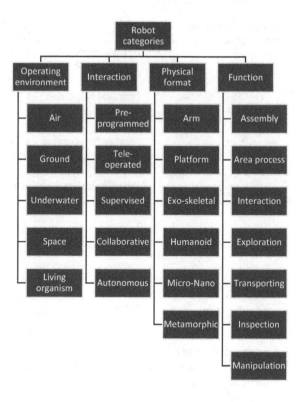

can operate within various groups of homogeneous units (swarm) or in interoperating collections of robots (ecosystems) that consist of individually operated units that share common objectives within a given operating environment. In the field of precision farming, teleoperation of an agricultural robotic system requires effective and efficient human–robot interaction. These interactions are expected to reach greater maturity and become more intuitive in the near future with the expected technological progress in gestural and spoken human–machine interaction.

Physical Format: The most common type of physical robot manifestation takes the form of jointed robot manipulators (arms) that allow robots to interact with their environment. They can be characterized into one of five major categories by their mechanical structure: (i) Cartesian, (ii) cylindrical (iii) spherical (polar), (iv) SCARA robots that all joints are parallel along Z-axis to allow full movement throughout a plane and (v) articulated robots that consist of three or more rotary joints. Robotic arms have been used typically for industrial applications that perform one or more tasks repeatedly with pre-determined movements. Data from sensors can enable execution of more complex actions, whereas the pairing with advanced analytics, machine vision and learning has been used recently to identify objects, act automatically and increase autonomy. Other physical forms of robots are exo-skeletal robots designed to help out with tough agricultural work, metamorphic with shape-changing capabilities, nano or micro robots (e.g. nanofertilizers and nanopesticides) and humanoid general-purpose robots.

Function: Several basic functions are usually combined by a robot to perform a specific task. Robots can perform assembly by joining parts together (e.g. welding with a fixing mechanism), they can be used for effectively carrying out surface processing or can inspect flat surfaces or an object of interest following pre-determined or random paths. They can interact with humans, other machines or robots involving physical contact or exchange of objects, grasping and manipulation. Robots are also capable of exploring an unknown or partially known environment with the goal to map the objects, people and resources that belong to it.

A wide range of technologies are typically integrated in every robotic system. Each technology enables specific functionalities that determine and characterize the whole system's operation. Depending on their use, robots are equipped with a selection of abilities including re-configurability, motion capabilities, abilities to handle objects, perceive the environment and act. A full list of robots' abilities is listed in the Table 1.2.

Table 1.2 List of robots' abilities and relevant description

Robot ability	Description
Configurability	The ability to be configured to perform one or more tasks
Adaptability	The ability to adapt itself to different work scenarios, different environments and conditions over long- or short-time scales
Interaction capability	The ability to interact both cognitively and physically, either with users, operators or other systems around it, including other robots
Dependability	The ability to perform its given tasks without systematic errors, thus specifying the level of trust that can be placed on the system to perform
Motion capability	The ability to move. Motion may be highly constrained (validation measures: precision of motion or repeatability) or unconstrained measured by the ability to move effectively in different media or between media
Manipulation ability	The ability to handle objects. End effectors can be fixed or specific to a task specifying the accuracy and repeatability of the manipulation
Perception ability	The ability to perceive its environment (e.g. detection of objects, obstacles, locations of interest). Also, ability to make informed and accurate deductions about the environment based on sensory data
Decisional autonomy	The ability to act autonomously. Nearly all systems have a degree of autonomy. It ranges from the simple motion of an assembly stopped by a sensor reading, to the ability to be self-sufficient in a complex environment
Cognitive ability	The ability to (i) interpret the task and environment (e.g. functions and interrelations between different objects) under environmental or task uncertainties, or both (ii) interpret human commands delivered in natural language or gestures and plan or execute tasks in response to these high-level commands

1.2.1.2 Robots in Agriculture

Prototype robotic systems and especially drones have been widely adopted in the agribusiness all over the world (Huuskonen and Oksanen 2018; Mogili and Deepak 2018). The main areas of application of agricultural robots include weed control, harvesting, planting seeds, harvesting, environmental monitoring and soil analysis, among others (Bakker et al. 2010; Bechar 2010; Bloch et al. 2018; Bochtis et al. 2011; McAllister et al. 2019).

Self-driving tractors: Self-driving tractors have achieved success in operating without a person inside the tractor itself. They can be programmed to observe their position independently, decide speed and avoid obstacles such as people, animals or objects in the field. They offer the possibility of autonomous seeding, planting and tillage, and are equipped with a variety of sensors such as radar, lasers and cameras. They allow path planning and adjustments depending on the needs of the operator, and multiple operations can be managed at the same time by multiple tractors in separate fields or in tandem in the same field. There are tractors that function with supervized autonomy where a leading tractor with an operator determines the path and the speed that are transmitted to the other tractors with vehicle-to-vehicle (V2V) technology and communication to exchange and share data. Driverless tractor technologies have also moved towards full autonomy or independent functioning, by integrating (i) laser or LiDAR sensing capabilities to detect and react to unforeseen obstacles, (ii) GPS positioning and radio feedback and (iii) automation software to manage the vehicle's path and control farming operations.

Fruit-picking agbots: Automated fruit picking in fields and smart machinery to remove the weeds that grow among crops have become reality with the latest advances in robot technology. The creation of such robots has become possible with the use and integration of multidisciplinary technologies including machine vision, electronics and mechanical engineering. Traditionally, robot technology was struggling with several challenges that involved, for example, the changing environment or the need to perform a variety of tasks required within a warehouse or field. Nowadays, high-powered image processing algorithms combined with low-cost but powerful sensors and hardware allow (i) efficient control of tasks that are not rigidly defined and (ii) handling of 'unknown' or partially known natural objects that are not always identical (typically come in a variety of sizes and geometries).

Drones: They have been used commercially since the 1980s, and are among the most promising technologies in agriculture with their potential to address several major challenges. Drone technology is expected to revolutionize the agricultural industry enabling planning and execution of operations based on real-time data gathering and processing (Table 1.3).

One of the most promising areas in agriculture where drones are expected to thrive involves fleets or swarms of autonomous drone 'actors' that act collectively and communicate and perform a variety of tasks allowing more concentrated inspection and treatment.

Table 1.3 Possible drone-powered solutions in agriculture

Operation	Enabling solution
Analysis	Drones can be used to generate precise 3-D maps for early soil analysis. *Possible applications*: management of seed planting patterns, irrigation and management of nitrogen status
Planting	Drone-planting systems can decrease planting costs by shooting pods with seeds and plant nutrients into the soil, providing the plant with all the nutrients necessary to sustain life
Spraying	Drones can scan the ground and spray the correct amount of liquid in real-time achieving increased efficiency with a reduction in the total spraying time and in the amount of chemicals penetrating into groundwater
Monitoring	Enhanced efficiency in crop monitoring is now achieved with the use of drones overcoming the drawbacks associated with satellite imagery (costly acquisition, quality suffered on cloudy days, images must be ordered in advance etc.). Today drones enable real-time and accurate monitoring of crop development and reveal production inefficiencies, enabling better crop management
Irrigation	Drones equipped with a variety of sensors (hyperspectral, multispectral or thermal) can identify which parts of a field are dry and therefore can be used to specify special irrigation strategies
Health assessment	Drone-carried devices can identify which plants reflect different amounts of green light and NIR light. They can also allow calculation of the normalized difference vegetation index, which describes the relative density and health of the crop. This information is essential to assess crop health and spot bacterial or fungal infections on trees

1.2.2 System Approach

The whole agri-food chain (as seen in Fig. 1.4) can be considered a functional succession of a number of dynamic, complex and logically interconnected operations or functions. Harvesting and handling operations can be seen as the vital link between the production activities and delivery of the crop from the time and place of harvest to the time and place of consumption with minimum losses and maximum efficiency. Agri-food losses can be either: (i) *quantitative* and related to physical substances that can be measured e.g. reduction in weight or volume and quality or (ii) *qualitative* and linked with the seed's productive potential, excessive respiration of products and

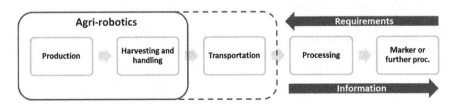

Fig. 1.4 The agri-robotics concept within the agri-chains system

food quality (e.g. total modification or decrease of food quality which makes it unfit for human consumption) that requires a different kind of evaluation.

Adopting a holistic approach, the agri-food chain consists of the production and subsequent processing and transformation of raw food products into consumer products for local and export markets. The main elements of the agri-food chain are given below:

A.   *Primary production*, which involves land preparation, irrigation, crop cultivation and harvesting and livestock farming.
B.   *Post-harvest and storage*, which is linked with a number of operations such as cleaning, sorting, cooling and packing the produce.
C.   *Transport and distribution*, when the raw product reaches the consumer or the processing site.
D.   *Processing*, which refers to the transformation of the raw product to meet the needs of the consumer. This stage may involve one or more of the following activities: drying, milling, grinding, pressing, shredding and de-husking.
E.   *Marketing,* as the final and decisive element in the post-harvest system before the end product is brought to the end users.

Robotics can be considered as a vital element at the first stages of the agri-chain systems where humans and robots have to work together (Bechar and Edan 2003). New levels of situational awareness have been achieved with the variety of wearables components and human-robot interfaces available today that are being used for action planning of robots and coordination between robots and humans. However, replacing manual work in the agricultural production domain with fully automated machine-based systems as seen in the industrial domain has proved to be a long way from being effectively applied yet. Hence, knowledge-based synergistic mechanization is needed as an intermediate option between the opposite extremes of manual and robotic labour. The agricultural sector is moving towards the generation of human–robot logistical synergies that will be applicable in operational environments of both outdoor in-field harvesting and handling operations and indoor storage operations prior to processing for high-value crops (HVC).

To achieve such a vision, a multidisciplinary approach is required with research areas such as engineering management, software engineering, information engineering, and so on, aiming to provide the technological framework, while other domains such as agriculture and business provide the operational framework of the system. The following technological and scientific challenges need to be addressed to meet the objectives of the rapidly evolving agricultural sector.

Activity modelling. The most critical operation within HVC production is manual fruit harvesting and handling. The worker-collector can be in different distinct states (e.g. crop collection, moving from one plant to the next, ascent–descent to the tree in the case of orchards, rest, and so on). In every case, his or her motion is continuous and gives a state transition time-series (e.g. position, orientation and other signals that differ depending on the conditions). In addition, discrete events related to label scanning, e.g. Radio-frequency identification (RFID) or barcodes from plants or other designated localization entities, crates, tools (e.g. ladder) follow different time distributions and sequences depending on the situation. Analogous operational features prevail in the indoor HVC storage environment. One of the challenges to be addressed is the modelling of HVC harvesting, handling and storage production processes as dynamic systems with discrete and continuous components. The main target of such approaches involves time scheduling of the tasks involved and the adoption of an object-oriented methodology for detailed modelling of tasks that are both continuous and discrete in HVC logistics.

Activity sensing of workers and robots. The HVC in-field and storage logistics are performed mainly by workers indicating that it is difficult to automate the data collection procedure. Many systems have been applied for mapping operations in orchards with encouraging results (Wulfsohn et al. 2012). A limited field data collection system, using mobile telephones (GSM–SMS), has been developed and tested with satisfactory results (Tseng et al. 2006). However, previous systems have been unable to overcome problems such as low-resolution maps, poor positioning accuracy, low grade process automation and incorrect or incomplete measurements. The advent of smart wearable devices (such as Artificial Intelligence empowered portable sensing devices) is expected to play a key role in collecting location data from GPS in the outdoor environment and attitude and heading reference systems (AHRS) for indoor environments (including roll, pitch and yaw, and heading information in a 3-D space). The data after processing will be able to reveal information relevant to workflow, and the activities of workers and robots to enhance the sensing capabilities of the systems used.

Activity and situation context recognition. The sensor data that relate to the operations, independently if the task is executed by a machine or a human, are either of a discrete or continuous nature (time series), whereas their interpretation depends on the type of task executed (context) and on the state of the entity involved. In the case of manual harvesting, signals from accelerometers can determine the type and speed of walking through extraction of appropriate features from time-series data. The recognition of context in a specific moment from different information sources has similar characteristics with sensor fusion, which is a special case of data fusion. The need for automated determination of traceability can be satisfied by employing methods for activity and situation context recognition where emphasis should be given to the development of techniques that can cope with heterogeneous data (e.g. advanced fusion, relevant learning algorithms and appropriate data pre-processing techniques such as principal component analysis (PCA) and independent component analysis (ICA)).

Robotic fleet logistics. Robotic fleets consist of either homogenous or heterogeneous (in terms of the executing task) cooperating units (e.g. transport platforms for the out-of-the orchard removal of the collected crops). Recent approaches have used the abstraction that the harvesters are the 'customers' in the vehicle routing problem with time windows (VRPTW) methodology (Bochtis and Sørensen 2009, 2010). They showed that operational planning problems involved in the logistics of such fleets of cooperating service units can be cast as instances of the VRPTW and consequently solved by relevant algorithmic approaches. As a real-life implementation, a project based on a coordinated team of peat moss harvesting tractor-robots in an open field was presented using on-board high-level controllers that performed (Johnson et al. 2009), for example path planning, mission execution and obstacle avoidance, while the coordination of these tractors was handled by a centralized component. In current research studies, planning as well as the information obtained from the recording or monitoring systems can, in most cases, be considered isolated, and not integrated as in a whole system. Dynamic versions of the aforementioned approach need to be developed to provide complete robotic solutions to fleet logistics for the coordination between service units of the fleet and between the fleet and workers.

Production information systems. Harvesting production data are of discrete and analogue nature (e.g. RFID data, acceleration pattern, pedometer data) that use different formats and coding. The lack of compatibility between hardware and software impedes communication and increases the complexity of information exchange. Current standards for information exchange in agriculture are based on ISO11783 or as it is commonly referred to as ISOBUS (a communication protocol for the agriculture industry) that attempts to standardize the communication between information systems and devices (e.g. a sensor). Current systems, like agroXML and PML (Physical Markup Language), aim to standardize data transfer. The use of semantic representation of the data collected together with appropriate data management strategies are essential to ensure and allow compatibility during wireless transmission and recording in a management information system.

Agri-robotics are expected to transform the whole agri-chain allowing optimization of HVC logistics operations through knowledge based human robot synergy, decreasing inputs, decreasing labour costs and assuring product quality. The information technology modules and infrastructures of the near future are expected to guarantee optimized use of resources and increased operational efficiency by integrating a variety of advanced functionalities such as robotic fleet management, logistical optimization of combined worker–robot operations and recognition of worker—machine activity.

1.2.3   A New Product Consideration

Considering the development of robotic solutions in agricultural production, several principles pertaining to the development of any new product or service have to be taken into account. These include the following (Bechar and Vigneault 2016, 2017):

- The new product or service must provide a solution that the customer really needs. To this end, in-depth knowledge of the user's requirements is a prerequisite for the generation of any successful business solution. Although various tools for the analysis of the user's requirements and the subsequent linking of these requirements with the design of service functionalities (such as quality function deployment, QFD (Carnevalli and Miguel 2008) has been implemented for the development of new technologically-advantaged products (Chan and Wu 2002).

In the agricultural domain there has been a limited number of cases of such a product development process (Sopegno et al. 2016).

- One major concern in a new product that executes a function is its reliability, which provides a measure of the confidence that the product will carry out the task without a total or partial failure. Reliability is a critical term for machinery that executes agricultural operations. To increase the reliability of a new system, all technologies implemented should be tested exhaustively, both individually and within other systems as well. This requires the implementation of technologies that already exist in operational environments and not of more advanced technologies still in a 'prototype' phase.

- The introduction of agri-robotics in agricultural production changes the whole production system either partially or as a whole depending of the amount of labour or conventional machinery replaced in the production chain (Marinoudi et al. 2019). When replacing existing solutions and existing production practises, the cost of the new solutions to the technology should be competitive with the existing ones (Lampridi et al. 2019a, b).

- The new product or service introduced in dynamic production systems, such as agriculture, should be a 'ready-to-use' system. An example of this in the agricultural operations domain is the fast and wide adoption of automatic guidance systems (Hameed et al. 2010; Holpp et al. 2013). These systems include the essential operating system for the functionality of precision agriculture related technologies and practices (Batte and Ehsani 2006; Russell and Norvig 2002). Plug-and-play standards are a critical requirement for the successful implementation of agri-robotics, and this concerns all of its human-interface components that should cope with the needs of a non-expert user.

- Because of the lack of a legal framework for the operation of robotic systems in open environments, such as in arable farming, safety and liability are of great concern and require the development and the subsequent introduction to the market of systems under the principle of 'humans-in-the-loop'.

References

Achillas C, Bochtis D, Aidonis D, Marinoudi V, Folinas D (2019) Voice-driven fleet management system for agricultural operations. Inf Process Agric 6(4):471–478. https://doi.org/10.1016/j.inpa.2019.03.001

Angelopoulou T, Tziolas N, Balafoutis A, Zalidis G, Bochtis D (2019) Remote sensing techniques for soil organic carbon estimation: a review. Remote Sensing 11(6):676. https://doi.org/10.3390/rs11060676

Bakker T, Van Asselt K, Bontsema J, Van Henten EJ (2010) Robotic weeding of a maize field based on navigation data of the tractor that performed the seeding. In: IFAC proceedings volumes (IFAC-PapersOnline), vol 3. https://doi.org/10.3182/20101206-3-jp-3009.00027

Batte MT, Ehsani MR (2006) The economics of precision guidance with auto-boom control for farmer-owned agricultural sprayers. Comput Electron Agric 53(1):28–44. https://doi.org/10.1016/J.COMPAG.2006.03.004

Bechar A (2010) Robotics in horticultural field production. Stewart Postharvest Rev 6(3):1–11. https://doi.org/10.2212/spr.2010.3.11

Bechar A, Edan Y (2003) Human-robot collaboration for improved target recognition of agricultural robots. Ind Robot: Int J 30(5):432–436. https://doi.org/10.1108/01439910310492194

Bechar A, Vigneault C (2016) Agricultural robots for field operations: Concepts and components. Biosyst Eng 149:94–111. https://doi.org/10.1016/J.BIOSYSTEMSENG.2016.06.014

Bechar A, Vigneault C (2017) Agricultural robots for field operations. Part 2: operations and systems. Biosyst Eng 153:110–128. https://doi.org/10.1016/J.BIOSYSTEMSENG.2016.11.004

Bloch V, Degani A, Bechar A (2018) A methodology of orchard architecture design for an optimal harvesting robot. Biosyst Eng 166:126–137. https://doi.org/10.1016/J.BIOSYSTEMSENG.2017.11.006

Bochtis DD, Sørensen CG (2009) The vehicle routing problem in field logistics part I. Biosyst Eng 104(4):447–457. https://doi.org/10.1016/j.biosystemseng.2009.09.003

Bochtis DD, Sørensen CG (2010) The vehicle routing problem in field logistics: Part II. Biosyst Eng 105(2):180–188. https://doi.org/10.1016/j.biosystemseng.2009.10.006

Bochtis DD, Sørensen CG, Jørgensen RN, Nørremark M, Hameed IA, Swain KC (2011) Robotic weed monitoring. Acta Agriculturae Scandinavica, Section B - Soil & Plant Science 61(3):202–208. https://doi.org/10.1080/09064711003796428

Bochtis DD, Sørensen CGC, Busato P (2014) Advances in agricultural machinery management: a review. Biosystems Engineering. Academic Press. https://doi.org/10.1016/j.biosystemseng.2014.07.012

Carnevalli JA, Miguel PC (2008) Review, analysis and classification of the literature on QFD—Types of research, difficulties and benefits. Int J Prod Econ 114(2):737–754. https://doi.org/10.1016/j.ijpe.2008.03.006

Chan L-K, Wu M-L (2002) Quality function deployment: A literature review. Eur J Oper Res 143(3):463–497. https://doi.org/10.1016/S0377-2217(02)00178-9

Godfray HCJ, Beddington JR, Crute IR, Haddad L, Lawrence D, Muir JF, ... Toulmin C (2010, February 12). Food security: the challenge of feeding 9 billion people. Science. https://doi.org/10.1126/science.1185383

Hameed IA, Bochtis DD, Sørensen CG, Nøremark M (2010) Automated generation of guidance lines for operational field planning. Biosyst Eng 107(4). https://doi.org/10.1016/j.biosystemseng.2010.09.001

Holpp M, Kroulik M, Kviz Z, Anken T, Sauter M, Hensel O (2013) Large-scale field evaluation of driving performance and ergonomic effects of satellite-based guidance systems. Biosyst Eng 116(2):190–197. https://doi.org/10.1016/j.biosystemseng.2013.07.018

Huuskonen J, Oksanen T (2018) Soil sampling with drones and augmented reality in precision agriculture. Comput Electron Agric 154:25–35. https://doi.org/10.1016/j.compag.2018.08.039

Johnson DA, Naffin DJ, Puhalla JS, Sanchez J, Wellington CK (2009) Development and implementation of a team of robotic tractors for autonomous peat moss harvesting. J Field Robot 26(6–7):549–571. https://doi.org/10.1002/rob.20297

Lampridi MG, Kateris D, Vasileiadis G, Marinoudi V, Pearson S, Sørensen ... CG, Bochtis D (2019a) A case-based economic assessment of robotics employment in precision arable farming. Agronomy 9(4):175. https://doi.org/10.3390/agronomy9040175

Lampridi MG, Sørensen CG, Bochtis D (2019b) Agricultural sustainability: a review of concepts and methods. Sustainability 11(18):5120. https://doi.org/10.3390/su11185120

Marinoudi V, Sørensen CG, Pearson S, Bochtis D (2019) Robotics and labour in agriculture. A context consideration. Biosyst Eng 184:111–121. https://doi.org/10.1016/J.BIOSYSTEMSENG.2019.06.013

McAllister W, Osipychev D, Davis A, Chowdhary G (2019) Agbots: weeding a field with a team of autonomous robots. Comput Electron Agric 163. https://doi.org/10.1016/j.compag.2019.05.036

Mogili UR, Deepak BBVL (2018) Review on application of drone systems in precision agriculture. In: Procedia computer science, vol 133, pp 502–509. Elsevier B.V. https://doi.org/10.1016/j.procs.2018.07.063

OECD-FAO (2012) OECD/FAO Agriculture Outlook 2012-2021. https://doi.org/10.1787/agr_out look-2012-en

Russell S, Norvig P (2002) Artificial intelligence: a modern approach, 2nd edn. Prentice Hall. https://doi.org/10.1017/S0269888900007724

Seyyedhasani H, Dvorak JS (2018) Dynamic rerouting of a fleet of vehicles in agricultural operations through a Dynamic Multiple Depot Vehicle Routing Problem representation. Biosyst Eng 171:63–77. https://doi.org/10.1016/J.BIOSYSTEMSENG.2018.04.003

Sopegno A, Calvo A, Berruto R, Busato P, Bocthis D (2016) A web mobile application for agricultural machinery cost analysis. Comput Electron Agric 130:158–168. https://doi.org/10.1016/J.COMPAG.2016.08.017

Sørensen CG, Bochtis DD (2010) Conceptual model of fleet management in agriculture. Biosyst Eng 105(1):41–50. https://doi.org/10.1016/j.biosystemseng.2009.09.009

Sørensen CG, Fountas S, Nash E, Pesonen L, Bochtis D, Pedersen … SM, Blackmore SB (2010a) Conceptual model of a future farm management information system. Comput Electron Agric 72(1):37–47. https://doi.org/10.1016/j.compag.2010.02.003

Sørensen CG, Jørgensen RN, Maagaard J, Bertelsen KK, Dalgaard L, Nørremark M (2010b) Conceptual and user-centric design guidelines for a plant nursing robot. Biosyst Eng 105(1):119–129. https://doi.org/10.1016/j.biosystemseng.2009.10.002

Sørensen CG, Pesonen L, Bochtis DD, Vougioukas SG, Suomi P (2011) Functional requirements for a future farm management information system. Comput Electron Agric 76(2):266–276. https://doi.org/10.1016/j.compag.2011.02.005

Tilman D, Balzer C, Hill J, Befort BL (2011) Global food demand and the sustainable intensification of agriculture. Proc Natl Acad Sci USA 108(50):20260–20264. https://doi.org/10.1073/pnas.1116437108

Tozer PR (2009) Uncertainty and investment in precision agriculture—Is it worth the money? Agric Syst 100(1):80–87. https://doi.org/10.1016/j.agsy.2009.02.001

Tseng CL, Jiang JA, Lee RG, Lu FM, Ouyang CS, Chen YS, Chang CH (2006) Feasibility study on application of GSM-SMS technology to field data acquisition. Comput Electron Agric 53(1):45–59. https://doi.org/10.1016/j.compag.2006.03.005

Wulfsohn D, Zamora FA, Téllez CP, Lagos IZ, García-Fiñana M (2012) Multilevel systematic sampling to estimate total fruit number for yield forecasts. Precis Agric 13(2):256–275. https://doi.org/10.1007/s11119-011-9245-2

Chapter 2
Agricultural Robotics for Precision Agriculture Tasks: Concepts and Principles

Avital Bechar

This chapter focuses on the principles, conditions and guidelines for agricultural robots to perform precision agricultural tasks, appraises the requirements of robotic systems, and presents associated concepts and characteristics of the complexities and types of precision agricultural tasks from a robotic perspective.

2.1 Introduction

Robots are perceptive machines that can be programmed to perform specific tasks, make decisions and act in real time. They are required in various fields that normally call for reductions in manpower and workload, and are best-suited for applications requiring repeatable accuracy and high yield under stable conditions (Holland and Nof 2007). However, they lack the capability to respond to ill-defined, unknown, changing, and unpredictable events (Moysiadis et al. 2020). Unlike industrial applications, which deal with simple, repetitive, well-defined and predetermined tasks, agricultural applications of automation and robotics require advanced technologies to deal with complex and highly variable environments and produce (Nof 2009). The technical feasibility of agricultural robots for a variety of agricultural tasks has been widely approved. Nevertheless, despite the tremendous amount of research, commercial applications of robots in complex agricultural environments are not yet available (Urrea and Munoz 2015). Such applications of robotics in uncontrolled field environments are still in the developmental stages (Bac et al. 2013). The main limiting factors lie in production inefficiencies and lack of economic justification. Development of an agricultural robot must include the creation of sophisticated,

A. Bechar (✉)
Institute of Agricultural Engineering, Volcani Institute, Rishon LeZion, Israel
e-mail: avital@volcani.agri.gov.il

© Springer Nature Switzerland AG 2021
A. Bechar (ed.), *Innovation in Agricultural Robotics for Precision Agriculture*,
Progress in Precision Agriculture,
https://doi.org/10.1007/978-3-030-77036-5_2

intelligent algorithms for sensing, planning and controlling to cope with the difficult, unstructured and dynamic aspects of agriculture (Bechar and Edan 2003).

In agriculture, the environment is very unstructured and demands the motion of robots unlike that of machines in a factory or of vehicles in a parking lot (Canning et al. 2004). It changes in time and space, with environmental conditions considered to be hostile and it requires mobile operation in 3-D changing tracks. The terrain, vegetation, landscape, visibility, illumination and other atmospheric conditions are not well defined; they vary continuously, have inherent uncertainty, and generate unpredictable and dynamic situations (Bechar and Vigneault 2017). Complexity increases when dealing with natural objects, such as fruits and leaves, because of the considerable variation in shape, texture, colour, size, orientation and position that in many cases cannot be determined a priori.

An example of variability in the agricultural environment is presented in Fig. 2.1, illustrating the variation and dynamics of the illumination levels in a bell pepper greenhouse that occur in a few hours and affect the visibility of the rows and the environment. Therefore, the task will require adaptive algorithms that could cope with the rapid changes in time.

From a robotic point of view, the world can be divided into four main domains, according to the structural characteristics of environments and objects: (a) the environment and the objects are structured, (b) the environment is unstructured and the objects are structured, (c) the environment is structured and the objects are unstructured and (d) the environment and objects are unstructured. Each robotic area such as industry, medicine, healthcare, and so on can be associated with one of the domains (Table 2.1). This illustrates the difference between the domains, their complexity and challenges. The agricultural domain is associated with the fourth, in which none

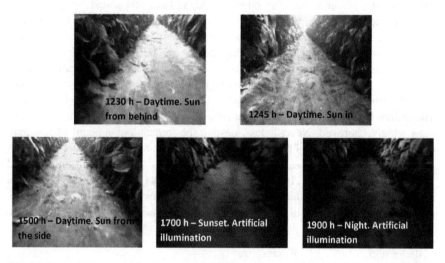

Fig. 2.1 Images of a pepper row in a greenhouse taken from a robotic platform at five different times in a day together with the illumination data (Dar et al. 2011)

Table 2.1 The four robotic domains (a variation on a table from Bechar and Vigneault (2016)

		Environment	
		Structured	Unstructured
Objects	Structured	Industrial/Service domains	Military/Space/Underwater domains
	Unstructured	Medical/Social domains	Agricultural domain

is structured and therefore, it is highly challenging to develop and commercialize. In such environments there are many situations in which autonomous robots fail because of the many unexpected events (Steinfeld 2004). This further complicates the robotic system and results in a system that is difficult and expensive to develop.

Figure 2.2 illustrates the difference in product weight distribution of agriculture and other domains. By quantifying the weight distribution of a specific product population with the coefficient of variation (CV, the standard deviation of the product population weight over the mean of the product population weight), the difference in product weights of different domains can be compared (Bechar and Vitner 2009). The analysis reveals that the CVs are small for metal, plastic and rubber products and vary between 0.01–0.05 and 0.07 to processed wood products. Small CV values represent a narrow population distribution and little variability. However, the CV value of agricultural products, in this case, flower cuttings have CVs that are one to two orders of magnitude larger (CV value of 0.34).

Growing and production processes in agriculture are complex, diverse, require intensive human labour and are usually unique to each crop. The process type and

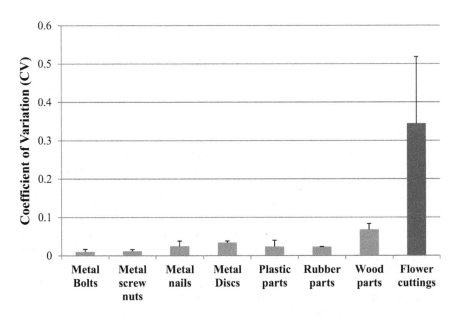

Fig. 2.2 Coefficient of variation (CV) of products in different domains

components are influenced by many factors, including: the crop characteristics and requirements, the geographical or geological environment, climatic and meteorological conditions (Tremblay et al. 2011), market demands, customers' requirements, and the farmer's capabilities and means. The technology, equipment and means that are required for a specific agricultural task involving any given crop and environment will not necessarily be applicable to another crop or in a different environment. The wide variety of agricultural systems and their diversity worldwide make it difficult to generalize the application of automation and control (Schueller 2006), therefore, more efficient agricultural practices are needed.

Agricultural productivity has increased markedly throughout the past 60 years, because of intensification, mechanization and automation. It is an important target for the application of various kinds of technologies designed to improve crop yields and other aspects of farming. In the 20th century, technological progress in developed countries has reduced the manpower for these activities by a factor of 80 (Ceres et al. 1998). Automation increases the productivity of agricultural machinery by increasing efficiency, reliability, quality, uniformity and precision, and reducing the need for human intervention. Although the evolution of technology and the transition to the digitized world of automation has triggered the introduction and use of autonomous robotic systems (Lampridi et al. 2019), one of the main limiting factors in the introduction of robotic systems to agriculture and precision agriculture is the high cost in applying such systems.

Autonomous robots in real-world, dynamic and unstructured environments still yield inadequate results (Bechar 2010), because of inherent uncertainties, unknown operational settings and unpredictable environmental conditions. Inadequacies of sensor technologies further impair the capabilities of autonomous robotics. Therefore, the promise of automatic and efficient autonomous operations has fallen short of expectations in unstructured and complex environments. Complexity increases with the involvement of natural objects, such as those encountered in medical and agricultural environments, because of the considerable variability in shape, texture, colour, size, orientation and position of such objects (Bechar et al. 2009). In addition, the product being dealt with is of relatively low cost, therefore the cost of the automated system must be low for it to be economically justified. Also, the seasonal nature of agriculture makes it difficult to achieve the high degree of utilization found in the manufacturing industries. The complex agricultural environment, combined with intensive production requires robust systems with short development time at low cost (Nof 2009).

The seasonality of agriculture makes it difficult to achieve the high level of utilization found in manufacturing. However, even if the technical and economic feasibility of most of the agricultural robotics applications is not reached in the near future using the existing knowledge and technologies, partial autonomy will add value to the machine long before autonomous robots are fully available. For many tasks, the Pareto principle applies. It claims that roughly 80% of a task is easy to adapt to robotics or automation, but the remaining 20% is difficult (Stentz et al. 2002). Therefore, by automating the easy parts of a task, one can reduce the required manual work by 80%. Furthermore, the development of partially autonomous robots is an excellent transitional path to developing and experimenting with software and hardware elements that will eventually be integrated into fully autonomous systems.

Precision agriculture (PA) was first introduce some four decades ago. The techniques and research in precision agriculture were conducted to align with four main objectives: to increase agricultural productivity, increase produce quality, reduce production costs and reduce environmental impact. Precision agriculture is the main beneficiary of the variability that defines the agricultural domain as discussed above. It aims to exploit the spatial variation using high resolution (up to a single plant level) decision-making and data collection to apply variable-rate operations to increase the total plot revenue and minimize the total cost. We can argue that If not for the variable nature of agriculture, precision agriculture would not be relevant. However, until recently, research in the fields of agricultural robotics and precision agriculture evolved along parallel paths with very little interaction, relation or reference between the two research fields.

Development of an agricultural robot to perform a precision agriculture task must start with development of integrated approaches and operation concepts of both robotics and precision agriculture and include the creation of sophisticated, intelligent algorithms for sensing, planning and control, and decision-making algorithms to cope with the difficult, unstructured and dynamic environment and the unique nature of precision agriculture tasks.

Referring to the three leading characteristics of the agricultural domain: the large degree of variation in the product, the level of structure in the environment and the systems costs, as dimensions in a domination space (Fig. 2.3). The agricultural domain is in the lower right area with high product variability, with poor structure level and low cost demand. It reveals the gaps that needs to be covered and the challenges of robotic systems for agriculture, and for precision agriculture in particular. Robotics is on the other side of the domination space dealing usually with little variation in the product, a well structured level in the environment and relatively large costs. The way to reduce the gap could be by developing concepts and approaches

Fig. 2.3 The domination space of the three dimensions: the product variability, the environment structure level in the environment and the cost. The blue line represents the gap robotics will need to cover and the challenges in this area

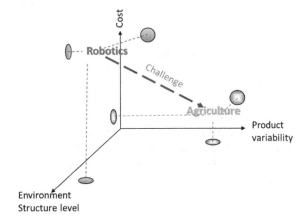

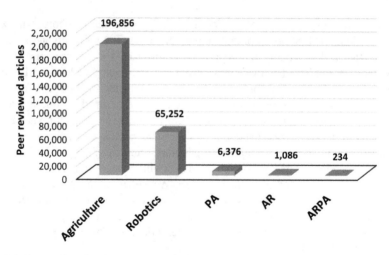

Fig. 2.4 Peer-reviewed articles on the main topic related to agricultural robotics for precision agriculture since 2015. *Source* Scopus, accessed in March 2020. PA—Precision Agriculture, AR—Agricultural Robots, ARPA—Agricultural Robots for Precision Agriculture

that are more suitable for precision agricultural tasks such as focusing on a specific task, and integrating a human operator into the robotic system, simplifying the robotic systems by creating robot teams and so on. These concepts are elaborated in Chaps. 7 and 8.

The relative research effort in the following areas: agriculture, robotics, precision agriculture (including precision farming and precision irrigation), agricultural robotics (AR) and robots for precision agriculture (ARPA) in the past five years is given in Fig. 2.4. It is based on peer-reviewed articles that have been published since 2015 according to Scopus. The annual average increase in the number of articles on PA, AR and ARPA topics is 15%, 20% and 15% respectively, and although 21% of the articles related to agricultural robots (AR) deals with precision agriculture tasks (ARPA), meaning it is an important field to the agricultural robotics community, only 3% of the articles related to precision agriculture topic (PA) were dedicated to agricultural robots.

Analysis of the frequencies of the main keywords in articles related to the ARPA topic revealed the most used keywords. They represent the areas that are investigated and provide an estimate of the directions that interest researchers working on robots for precision agriculture. Figure 2.5 shows the 'normalized frequencies' of the main keywords. 'Normalized frequency' is the number of times that a keyword appears divided by the number of articles on the same topic, i.e., for the keyword 'weed' (with all its derivatives: weed, seeding, etc.), the normalized frequency value is 14.2. This means that on average this keyword appears in 14.2% of the articles related to the ARPA field and probably deal with the precision agriculture task of weed detection, distribution or weeding. Based on this analysis, it seems that the main keywords related to 'agricultural operations' in the ARPA field are weed, harvest,

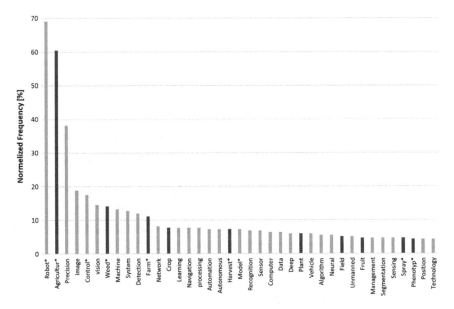

Fig. 2.5 The normalized frequencies of the main keywords used in ARPA articles in the past five years. *Source* Scopus, accessed in March 2020. The green bars represents keywords related to agriculture (crops, operations, etc.). A keyword with an asterisk represents all derivatives of the keyword

fruit, spraying and phenotyping which appear on average in 14.2, 7.3, 4.7, 4.7 and 4.3% of the articles respectively. The keywords related to 'agricultural environment' are farm, field, crop, plant and fruit which appear on average in 11.2, 5.2, 7.7, 6 and 4.7% of the articles respectively.

2.2 Basic Guidelines and Conditions for Applying Robots in Precision Agricultural Tasks

Much research has been carried out on agricultural robotics in the past 40 years. Almost all of them did not reach the commercialization stage. The main causes for incompletion were the extensive costs of the robots developed, inability to execute the required agricultural task, lack of robustness of the system, and inability to reproduce the same task successfully in slightly different contexts or to satisfy operational or economic aspects of the agricultural task. In addition, most approaches were imported from the industrial domain (Vidoni et al. 2015) and did not fit to the tasks in hand. All the effort conducted so far has enabled the formulation of guidelines and definitions of the basic conditions required for development of agricultural robots (Bechar and Vigneault 2016) with modification to precision agriculture. The development and application of robots for precision agricultural tasks has to comply with the following five guidelines:

1. The farmer's requirements for manipulating specific produce must be considered first.
2. The precision agricultural task and its components must be feasible using the existing technology and the required complexity.
3. The required spatial and temporal resolution must be feasible by the robotic system and synchronized with other tasks in the process chain.
4. The cost of the robotic system solution must be less than the expected revenue. It is not necessary that it should be the most profitable alternative.
5. The robotic system developed must have an added value for the performance of the precision agriculture task or for other tasks in that process.

In most cases, the use of robots to perform precision agriculture tasks is achievable if at least one of the following conditions is met:

a. The cost of utilizing robotics is less than the cost of any concurrent methods.
b. The use of robotics enables increasing farm production capability, produce, profit and survivability under competitive market conditions.
c. The use of robotics improves the quality and uniformity of the produce.
d. The use of robotics minimizes the uncertainty and variation in growing and production processes.
e. The use of robotic systems enables the farmer to make decisions and act at greater temporal or spatial resolution compared to the current system to achieve optimization in the growing and production stages in an equivalent manner to 'lean manufacturing' in industry.
f. The use of robotic systems enables an increase in the quality of service or information.
g. The robotic system is able to perform specific tasks that are defined as hazardous or that cannot be performed manually.

2.3 Principles and Classification of Precision Agricultural Tasks for Robotic Applications

Much research has been conducted worldwide in the field of robots for precision agriculture recently (Conesa-Munoz et al. 2015; Bhimanpallewar and Narasingarao 2020; Raja et al. 2020a, b; Sai et al. 2019; Thayer et al. 2020; Ünal et al. 2020). This research has demonstrated the technical feasibility of agricultural robots for a variety of crops, precision agriculture tasks and robotic abilities. However, automation solutions have not yet been commercially implemented successfully for field operations and only a few developments have been adopted and put into practice (Xiang et al. 2014). Incompatibility between the robotic system designed and the precision agriculture task led to production inefficiencies, long cycle times and delays, low detection rates (Zhao et al. 2016) and the inability to perform the necessary PA tasks satisfactorily. The unstructured nature of agricultural environments generates stochastic task requirements and the live and fragile plant and produce make features of the agricultural task quite different from industrial applications that work with inorganic products.

Robots for precision agriculture tasks comprise numerous sub-systems and devices that enable them to operate and perform their tasks. These sub-systems and devices deal with path planning, navigation or guidance abilities (Carpio et al. 2020, Zaidner and Shapiro 2016), mobility, steering and control (Lipinski et al. 2016), sensing, manipulators or similar functional devices (Mann et al. 2014), end effectors, control, decision-support systems to manage individual or simultaneous unexpected events, and some level of autonomy (van Henten et al. 2013). Robots for precision agriculture are generally designed to execute a specific agricultural task, such as specific spraying (Asaei et al. 2019), selective weeding (Wu et al. 2020b), disease monitoring (Kerkech et al. 2020, Liang et al. 2020), selective pruning (Bechar et al. 2014), and so on. These are considered to be the 'main tasks' to be performed by the robotic system. To execute the 'main task' successfully, the robotic system must perform several 'supporting tasks', such as localization and navigation, detection of the object to treat, etc. Information and commands are transferred between the 'supporting tasks' and the 'main task'. Each 'supporting task' controls one or several sub-systems and devices, and a sub-system or device may serve several 'supporting tasks' (Fig. 2.6). For instance, in developing a disease monitoring robot (Schor et al. 2016a), the 'main task' is disease monitoring, the robotic system needs to be able to perform the 'supporting tasks' of self-localization, trajectory planning, steering and

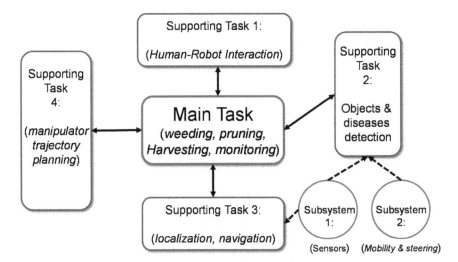

Fig. 2.6 Structure of task sub-systems in an agricultural robot. Solid arrows represent commands, data and information transfer; dashed arrows represent conceptual connections. The writing in the parentheses are examples for agricultural robot 'main tasks', 'supporting tasks' and subsystems (Bechar and Vigneault 2016)

navigating in the plot from its actual location to the next sampling location, collaborating with a human operator or interacting with a human presence, other robots or unexpected objects on the path and to modify its trajectory planning as necessary. Nguyen et al. (2013) developed and implemented a framework for motion and hierarchical task planning for an apple harvesting robot, Bechar et al. (2009) developed a methodology for melon detection by a human–robot system to be used by a melon harvesting robot and Ceres et al. (1998) developed and implemented a framework for a human integrated citrus harvesting robot. A framework for agricultural and forestry robots was developed by Hellstrom and Ringdahl (2013).

Further investigation of the precision agriculture task characteristics, i.e. the 'main task' to execute in the robotic framework, reveals that it can be classified into a three-level scale based on the task complexity. The task complexity can be defined by the level of robot–plant interaction, whereas higher level represents greater challenges. The lower level of complexity of the robot–plant interaction requires no physical contact between the robot and the plant. At this level, the precision agriculture tasks are involved mainly in (i) data collection using visual and other sensors (elaborated in Chap. 3), e.g. early detection of diseases and pests, abiotic stress diagnostics and identification of anomalies (Sanchez et al. 2020; Freitas et al. 2020), (ii) transportation of produce, materials and tools between different locations of the farm (Guzman et al. 2016) and (iii) remote material application such as variable-rate fertilizer application, selective and specific spraying, and so on (see more in Chap. 6). The middle level of complexity requires physical contact between the robot and the plant but no handling of produce, materials or plant parts. Typical precision agriculture tasks at this level are selective mechanical weeding (Tillett et al. 2008) that will physically damage the weed but does not collect or handle it, seedling, fruit thinning, and branch pruning that removes fruitlets and branches, etc. The third level of complexity of the robot–plant interaction and the most challenging one requires both physical contact between the robot and the plant and handling of produce, materials or plant parts. Among the tasks at this complexity level would be fruit picking, harvesting of leaf crops, which require precise operation, decision-making and handling the produce without impairing it or reducing its quality. Transplanting of plants and trees, transferring of pots (with plants) in plant nurseries, and so on.

In addition, since the main objectives of precision agriculture tasks are either to collect data, analyse it, make decisions or act accordingly at a higher resolution, up to the plant level, precision agriculture tasks can be defined and classified according to three phases or stages concerning the operation of agricultural robots in executing the 'main task'. The first stage of a PA 'main task' deals with data collection. Representative tasks in this stage are high spatial and temporal resolution monitoring of climate and environmental conditions, soil sampling (Lukowska et al. 2019; Schnug et al. 1998) for nutrients, pests and bacteria, visual and acoustic monitoring (Finkelshtain et al. 2017; Schor et al. 2016b) of anomalies, biotic and abiotic stresses (Wang et al. 2019), yield and plant conditions. The second stage is attributed to decision-making, optimization and decision-support processes. Characterizing PA tasks at this stage are irrigation management interfaces, classification tasks, planning of farm processes and so on. The third stage relates to tasks that require action or

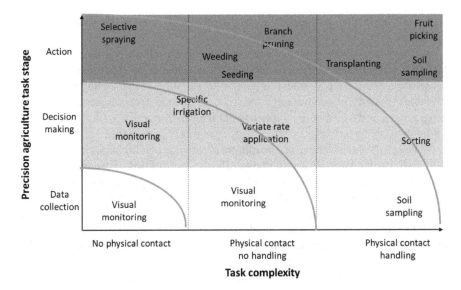

Fig. 2.7 The precision agriculture task classification space based on the task complexity level and the precision agriculture stage. The location of several different tasks in this space can demonstrate the challenge level. The blue lines represent equal level values of challenges and research and development effort of robotics in performing a precision agricultural task

physical performance such as specific spraying, transplanting and seeding (Gao et al. 2016; Bhimanpallewar and Narasingarao 2020), weed control (Wu et al. 2020a; Raja et al. 2020a), fruit picking and harvesting (Bloch et al. 2018), etc.

Combining the two classifications of precision agriculture tasks discussed above and creating a task classification space (Fig. 2.7), can enable us to position a specific task and to estimate the challenge level, and the required research and development effort in designing a robot to perform that task (Fig. 2.7). In this analysis the two classification dimensions have a similar influence on the challenge level. The challenge level of a specific task can be evaluated qualitatively by the magnitude of the distance between the task location to the origin of the axes.

2.4 Conclusions

Research, developments and evaluations of robots to perform precision agriculture tasks are very diverse in terms of objectives, structures, techniques and components. In this context, it is difficult to compare different robots and to transfer developed technology from one task to another. The limiting factors for the development of such systems are unique to each robotic system and precision agriculture task. In this chapter, an investigation of the characteristics of precision agriculture tasks was conducted and an evaluation platform between different systems and tasks was created.

Research and development of robotic systems to perform precision agriculture tasks need to follow several steps. First, investigate and study the nature of the task, the process and the environment in relation to variation in the leading variables to evaluate the feasibility of the suggested solution. Second, technologies and methodologies must be developed or modified to fit high variable situations and to overcome difficult problems such as the continuously changing conditions, the variability of the produce and the environment, and hostile environmental conditions such as vibration, dust, extreme temperature and humidity. Third, Identification of processes or tasks that can be 'robotized', evaluation of the overall task complexity and the precision agriculture stage. Fourth, evaluation of the challenge level and the required research and development effort for such a system and tasks. For very complex tasks, a high challenge level or large research and development effort, possible solutions to overcoming this problem might be agronomic modifications or a human integration. Fifth, to investigate if the solution presented complies with the guidelines and conditions discussed in Sect. 2.2. Finally, agricultural robotic systems should be developed only from tasks and processes where other solutions, such as mechanics or automation, cannot exist or that robotics has a diminishing marginal utility with use of them.

The robots that are to be used for precision agriculture tasks must recognize and understand the physical properties of each specific object, and must be able to work under different and dynamic environmental conditions in fields, or in controlled environments. Therefore, they need sensing systems that can work under variable conditions, as well as specialized manipulators and end-effectors. The environmental conditions are occasionally so severe with regard to high temperature, humidity, dust and or rain that electrical circuit and material corrosion problems can be a major concern. These conditions must be taken into consideration when designing robotic systems for precision agriculture tasks. In this sense, development and application of robots for precision agriculture tasks is an iterative process.

References

Asaei H, Jafari A, Loghavi M (2019) Site-specific orchard sprayer equipped with machine vision for chemical usage management. Comput Electron Agric 162:431–439

Bac CW, Hemming J, van Henten EJ (2013) Robust pixel-based classification of obstacles for robotic harvesting of sweet-pepper. Comput Electron Agric 96:148–162

Bechar A (2010) Robotics in horticultural field production. Stewart Postharvest Rev 6:1–11

Bechar A, Bloch V, Finkelshtain R, Levi S, Hoffman A, Egozi H, Schmilovitch Z (2014) Visual servoing methodology for selective tree pruning by human-robot collaborative system. AgEng 2014, Zurich, Switzerland

Bechar A, Edan Y (2003) Human-robot collaboration for improved target recognition of agricultural robots. Ind Robot 30:432–436

Bechar A, Meyer J, Edan Y (2009) An objective function to evaluate performance of human-robot collaboration in target recognition tasks. IEEE Trans Syst Man Cybern Part C-Appl Rev 39:611–620

Bechar A, Vigneault C (2016) Agricultural robots for field operations: concepts and components. Biosyst Eng 149:94–111

Bechar A, Vigneault C (2017) Agricultural robots for field operations. Part 2: operations and systems. Biosyst Eng 153:110–128

Bechar A, Vitner G (2009) A weight coefficient of variation based mathematical model to support the production of 'packages labelled by count' in agriculture. Biosyst Eng 104:362–368

Bhimanpallewar RN, Narasingarao MR (2020) AgriRobot: Implementation and evaluation of an automatic robot for seeding and fertiliser microdosing in precision agriculture. Int J Agric Resour Governan Ecol 16:33–50

Bloch V, Degani A, Bechar A (2018) A methodology of orchard architecture design for an optimal harvesting robot. Biosyst Eng 166:126–137

Canning JR, Edwards DB, Anderson MJ (2004) Development of a fuzzy logic controller for autonomous forest path navigation. Trans ASAE 47:301–310

Carpio RF, Potena C, Maiolini J, Ulivi G, Rossello NB, Garone E, Gasparri A (2020) A Navigation architecture for ackermann vehicles in precision farming. IEEE Robot Autom Lett 5:1103–1110

Ceres R, Pons FL, Jimenez AR, Martin FM, Calderon L (1998) Design and implementation of an aided fruit-harvesting robot (Agribot). Ind Robot 25:337+

Conesa-Munoz J, Gonzalez-De-soto M, Gonzalez-De-santos P, Ribeiro A (2015) Distributed multi-level supervision to effectively monitor the operations of a fleet of autonomous vehicles in agricultural tasks. Sensors 15:5402–5428

Dar I, Edan Y, Bechar A (2011) An adaptive path classification algorithm for a pepper greenhouse sprayer. American Society of Agricultural and Biological Engineers Annual International Meeting 2011,Louisville, KY. American Society of Agricultural and Biological Engineers, pp 288–302

Finkelshtain R, Bechar A, Yovel Y, Kósa G (2017) Investigation and analysis of an ultrasonic sensor for specific yield assessment and greenhouse features identification. Precision Agric 18:916–931

Freitas H, Faical BS, Silva A, Ueyama J (2020) Use of UAVs for an efficient capsule distribution and smart path planning for biological pest control. Comput Electron Agric 173

Gao GH, Feng TX, Yang H, Li F (2016) Development and optimization of end-effector for extraction of potted anthurium seedlings during transplanting. Appl Eng Agric 32:37–46

Guzman R, Navarro R, Beneto M, Carbonell D (2016) Robotnik-professional service robotics applications with ROS. In: Koubaa A (ed) Robot operating system

Hellstrom T, Ringdahl O (2013) A software framework for agricultural and forestry robots. Ind Robot: Int J 40:20–26

Holland SW, Nof SY (2007) Emerging trends and industry needs. Handbook of Industrial Robotics. Wiley, New York

Kerkech M, Hafiane A, Canals R (2020) Vine disease detection in UAV multispectral images using optimized image registration and deep learning segmentation approach. Comput Electron Agric 174

Lampridi MG, Kateris D, Vasileiadis G, Marinoudi V, Pearson S, Sørensen CG, Balafoutis A, Bochtis D (2019) A case-based economic assessment of robotics employment in precision arable farming. Agronomy 9

Liang H, He J, Lei JJ (2020) Monitoring of corn canopy blight disease based on UAV hyperspectral method. Spectroscopy Spectral Anal 40:1965–1972

Lipinski AJ, Markowski P, Lipinski S, Pyra P (2016) Precision of tractor operations with soil cultivation implements using manual and automatic steering modes. Biosyst Eng 145:22–28

Lukowska A, Tomaszuk P, Dzierzek K, Magnuszewski L (2019) Soil sampling mobile platform for Agriculture 4.0

Mann MP, Rubinstein D, Shmulevich I, Linker R, Zion B (2014) Motion planning of a mobile cartesian manipulator for optimal harvesting of 2-D crops. Trans ASABE 57:283–295

Moysiadis V, Tsolakis N, Katikaridis D, Sørensen CG, Pearson S, Bochtis D (2020) Mobile robotics in agricultural operations: a narrative review on planning aspects. Appl Sci (Switzerland), 10

Nguyen TT, Kayacan E, de Baedemaeker J, Saeys W (2013) Task and motion planning for apple harvesting robot. IFAC Proceedings Volumes 46:247–252

NOF SY (ed) (2009) Handbook of Automation. Springer

Raja R, Nguyen TT, Slaughter DC, Fennimore SA (2020a) Real-time weed-crop classification and localisation technique for robotic weed control in lettuce. Biosyst Eng 192:257–274

Raja R, Nguyen TT, Vuong VL, Slaughter DC, Fennimore SA (2020b) RTD-SEPs: real-time detection of stem emerging points and classification of crop-weed for robotic weed control in producing tomato. Biosyst Eng 195:152–171

Sai KK, Satyanarayana P, Hussain MA, Suman M (2019) A real time precision monitoring and detection of rice plant diseases by using internet of things (IoT) based robotics approach. Int J Innov Technol Exploring Eng 8:403–408

Sanchez L, Pant S, Mandadi K, Kurouski D (2020) Raman spectroscopy vs quantitative polymerase chain reaction in early stage Huanglongbing diagnostics. Sci Rep 10

Schnug E, Panten K, Haneklaus S (1998) Sampling and nutrient recommendations—The future. Commun Soil Sci Plant Anal 29:1455–1462

Schor N, Bechar A, Ignat T, Dombrovsky A, Elad Y, Berman S (2016a) Robotic disease detection in greenhouses: combined detection of powdery mildew and tomato spotted wilt virus. IEEE Robot Autom Lett 1:354–360

Schor N, Berman S, Dombrovsky A, Elad Y, Ignat T, Bechar A (2016b) Development of a robotic detection system for greenhouse pepper plant diseases. Precision Agriculture (In press)

Schueller JK (2006) CIGR handbook of agricultural engineering, CIGR—The International Commission of Agricultural Engineering

Steinfeld A (2004) Interface lessons for fully and semi-autonomous mobile robots. In: IEEE international conference on robotics and automation, pp 2752–2757

Stentz A, Dima C, Wellington C, Herman H, Stager D (2002) A system for semi-autonomous tractor operations. Auton Robots 13:87–104

Thayer TC, Vougioukas S, Goldberg K, Carpin S (2020) Multirobot routing algorithms for robots operating in vineyards. IEEE Trans Autom Sci Eng

Tillett ND, Hague T, Grundy AC, Dedousis AP (2008) Mechanical within-row weed control for transplanted crops using computer vision. Biosyst Eng 99:171–178

Tremblay N, Fallon E, Ziadi N (2011) Sensing of crop nitrogen status: opportunities, tools, limitations, and supporting information requirements. Horttechnology 21:274–281

Ünal İ, Kabaş Ö, Sözer S (2020) Real-time electrical resistivity measurement and mapping platform of the soils with an autonomous robot for precision farming applications. Sensors (Switzerland) 20

Urrea C, Munoz J (2015) Path tracking of mobile robot in crops. J Intell Rob Syst 80:193–205

van Henten EJ, Bac CW, Hemming J, Edan Y (2013) Robotics in protected cultivation. IFAC Proceedings Volumes 46:170–177

Vidoni R, Bietresato M, Gasparetto A, Mazzetto F (2015) Evaluation and stability comparison of different vehicle configurations for robotic agricultural operations on side-slopes. Biosyst Eng 129:197–211

Wang D, Vinson R, Holmes M, Seibel G, Bechar A, Nof S, Tao Y (2019) Early detection of tomato spotted wilt virus by hyperspectral imaging and outlier removal auxiliary classifier generative adversarial nets (OR-AC-GAN). Sci Rep 9

Wu X, Aravecchia S, Lottes P, Stachniss C, Pradalier C (2020) Robotic weed control using automated weed and crop classification. J Field Robot 37:322–340

Xiang R, Jiang H, Ying Y (2014) Recognition of clustered tomatoes based on binocular stereo vision. Comput Electron Agric 106:75–90

Zaidner G, Shapiro A (2016) A novel data fusion algorithm for low-cost localisation and navigation of autonomous vineyard sprayer robots. Biosyst Eng 146:133–148

Zhao Y, Gong L, Huang Y, Liu C (2016) Robust tomato recognition for robotic harvesting using feature images fusion. Sensors 16

Chapter 3
Agricultural Robotic Sensors for Crop and Environmental Modelling

Alexandre Escolà, Fernando Auat Cheein, and Joan R. Rosell-Polo

3.1 Introduction

The first stage in the Precision Agriculture (PA) cycle entails data acquisition. This chapter deals with sensors used in conjunction with ground robotic platforms or terrestrial service units involved in scouting operations to model crops and their surrounding environment in PA and in Precision Fructiculture. In the following sections, various sensors and techniques used on ground robotic platforms and for geometric and structural modelling of crops and their environment are presented. A subsequent stage in the PA cycle covers data processing and information extraction to support farmers in taking management decisions. Sensor data and derived information can be used both for navigation in service units and for acquiring, featuring and classifying agricultural scenes. A description of the most used sensors, their functioning principles and processing techniques is provided, as well as real applications in case studies.

3.2 Classification of Sensors

In robotics, there is a wide variety of sensors focused mainly on the interaction of the robot with the environment and sensors that provide information on the internal status of the robot. In agriculture, a service unit (i.e. an automated agricultural machine)

A. Escolà (✉) · J. R. Rosell-Polo
Research Group on AgroICT & Precision Agriculture, Department of Agricultural and Forest Engineering, Universitat de Lleida – Agrotecnio-CERCA Centre, Lleida, Catalonia, Spain
e-mail: alex.escola@udl.cat

F. A. Cheein
Department of Electronic Engineering, Universidad Técnica Federico Santa María, Valparaíso, Chile

© Springer Nature Switzerland AG 2021
A. Bechar (ed.), *Innovation in Agricultural Robotics for Precision Agriculture*,
Progress in Precision Agriculture,
https://doi.org/10.1007/978-3-030-77036-5_3

is usually equipped with enough sensors to guarantee the successful execution of a specific agricultural task. Therefore, sensors are related to the task being executed and the market offers a wide range of sensory solutions. Nevertheless, sensors in robotics can be classified according to the way they interact with the environment and the variable they measure. Thus, two main classifications are possible: (1) proprioceptive or exteroceptive sensors and (2) active or passive sensors. One classification group is not necessarily orthogonal to the other; a sensor can be active or passive and proprioceptive or exteroceptive. Table 3.1 summarizes the classification of sensors in robotics applied to agriculture. Briefly:

- **Proprioceptive sensors** measure internal variables of the service unit, such as wheel speed, battery charging state, internal temperature and dead-reckoning, among others. Proprioceptive sensors are more related to the mechatronic design of the service unit.
- **Exteroceptive sensors** acquire information from the surrounding environment, such as RGB (red, green, blue) cameras and ultrasonic sensors, light detection and ranging (LiDAR)-based sensors, time-of-flight (ToF)-based cameras, among others. These sensors provide information from crops and agricultural scenes.

  According to the way sensors interact with the environment:

- **Active sensors** interact with the environment through energy interchanging. That means the sensor uses its own power source to emit energy to the environment and records the returning energy. This first category comprises LiDAR sensors, ultrasonic sensors and ToF cameras, as examples, since they use their own light or ultrasound source.
- **Passive sensors** record energy reflected by the environment emitted by an external source. The RGB cameras are passive sensors; they capture light from natural or artificial sources, reflected by the surrounding objects.

3.3 Sensing Principles

In the previous section, sensors used most by robots in agricultural applications were listed. However, the different sensors have different technological principles and provide different data. Users should take into account the different operational constraints and capabilities offered by each sensor. This section provides the reader with the basic technological and operational principles of the most common sensors used by robots in agriculture.

Table 3.1 List of the most commonly used sensors in agriculture and their classification

General classification	Sensor principle or system	Proprioceptive (PC) or Exteroceptive (EC)	Passive (P) or Active (A)
Tactile sensors	Contact switches	EC	P
Wheel sensors	Encoders and potentiometers	PC	A or P
Orientation, attitude and heading sensors	Gyroscopes, accelerometers, inertial measurement units (IMUs)	PC or EC	A or P
Landmarks	GPS/GNSS, radio frequency, ultrasound, reflecting landmarks	EC	A or P
Range sensors	LiDAR, radar, ultrasound, infrared proximity diodes	EC	A
Velocity and motion sensors	Radar	EC	A
Machine vision systems	CCD/CMOS in RGB cameras	EC	P

3.3.1 RGB Cameras

An RGB camera is a low cost solution for many applications in agricultural scenarios, crop detection and classification, localization and mapping, safe navigation of service units and weed detection, among others (Assirelli et al. 2015; Berge et al. 2012; Bossu et al. 2009; Cheein et al. 2011; Tang et al. 2016; Tillett et al. 2002). However, such cameras are very sensitive to lighting conditions; they may require calibration of the focal centre, compensation for deformation of the lens and, moreover, they do not provide depth information directly unless stereovision or other multi-camera arrangements are used (Bietresato et al. 2016; Zhai et al. 2016). When single RGB cameras are used in service units, it should be noted that they cannot be the only exteroceptive sensor because its sensitivity to lighting conditions might saturate the sensor. Nevertheless, RGB cameras in agricultural scenarios are used successfully for weed detection, fruit classification, vertical farming and terrain classification, always under controlled lighting conditions (Cho et al. 2002).

Depending on the manufacturer, digital RGB cameras can operate at rates up to 64 frames per second (FPS) using the classical USB 2.0 protocol or up to 300 FPS or even higher using USB 3.0 protocol. It is worth mentioning that such a data bandwidth is greatly affected in the transmission of the information by the operating system of the acquisition controller, the data processing and the expected resolution of the image.

3.3.2   Ultrasonic Range Sensors

Ultrasonic range sensors are one of the sensors most used in robotics. They are exteroceptive and active sensors used either for object or obstacle detection or for detection and ranging. Their output is usually a digital signal when used for detection and an analogue voltage signal when designed for ranging applications. The sensor operation is based on the time-of-flight (ToF) principle.

Ultrasonic sensors are used for obstacle detection and navigation in robotics (Gutierrez–Osuna et al. 1998; Ortega and Camacho 1996). In addition, they are also used to detect and characterize electronically crop canopies when mounted on service units (see Sect. 3.3.1). Regardless of the application, special care should be taken with the mounting position of the sensors, the ultrasound cone angle and possible interference with adjacent sensors. Ultrasonic sensors should be mounted in a position that maximizes the energy of the possible echoes. When used in the field with natural targets, ultrasonic sensors might not follow the manufacturer's patterns strictly in terms of detection cone shape and distances. An in situ configuration (i.e. maximum range and attenuation factor) and calibration is recommended. Additionally, when several sensors operate within the same area, interference might occur. Preventive measures to mitigate interference might be to increase the distance between active sensors, synchronize the sensor operation to avoid close sensors being operated simultaneously and isolate the sound pathways to avoid side waves reaching the sensor membranes.

3.3.3   LiDAR Sensors

Light detection and ranging sensors have been used extensively in forestry and civil engineering or architecture applications for a long time (Maclean and Krabill 1986; Mukupa et al. 2016; Yan et al. 2015) whereas their use in agriculture goes back to just 20 years approximately. Pioneer LiDAR applications in agriculture were mostly related to the improvement and optimization of the application of plant protection products (PPP) in tree crops. In such situations, LiDAR sensors were initially used either for real-time control of PPP flow rates applied by sprayers in an on-the-go basis (Escolà et al. 2013; Llorens et al. 2011; Wangler et al. 1993; Wei and Salyani 2004) or to characterize the crop's geometric and structural properties to derive PPP dose recommendations (Walklate 1989; Walklate et al. 1997). However, other agricultural applications of LiDAR sensors soon appeared that aimed to improve other agricultural tasks. Research on the use of terrestrial laser scanners (TLS) and mobile terrestrial laser scanners (MTLS) has been done in irrigation, fertilizer application, pruning, tree training and weed control based on the geometric and structural properties of plants and crops (Andújar et al. 2013; Arnó et al. 2013; Palacín et al. 2007; Rosell et al. 2009; Rosell and Sanz 2012; Tumbo et al. 2002).

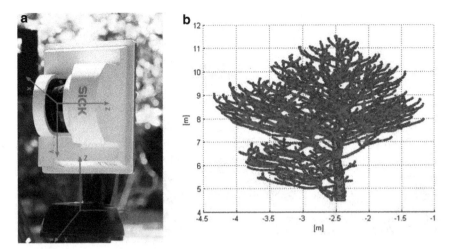

Fig. 3.1 Terrestrial laser scanner used in morphological modelling of trees. The LiDAR sensor used (**a**) was mounted on a robotized tripod and the system was used to model a cherry tree, *Prunus avium* L., (**b**) in the field (adapted from Pfeiffer et al. 2018)

Formerly, 3-D TLS were to be used under stationary conditions, but research and also commercial applications soon demanded new solutions and they are currently available for mobile use (MTLS). In this regard, both 2-D and 3-D TLS are used in urban and forestry applications whereas 2-D TLS is mostly used in agricultural applications (although the use of 3-D TLS in agriculture is growing rapidly). For example, Fig. 3.1a shows a ground SICK LiDAR sensor and Fig. 3.1b its application in morphological characterization of a cherry tree (*Prunus avium* L.) in Chile.

Regardless of the field of view, the principle of most LiDAR sensors consists of a laser beam emitting source and a receiving photodetector. The primary measuring configuration is one-dimensional: the emitted laser beam affects a delimited area of the target and a fraction of the backscattered light is captured and converted into a voltage signal by the photodetector; 2-D LiDAR sensors scan the targets in a plane by modifying the direction of the laser beam through an internal spinning mirror (Fig. 3.2a); 3-D LiDAR sensors add the third dimension by rotating the 2-D LiDAR around the remaining coordinate axis or by adding a second spinning mirror. Recently, a new generation of solid state LiDAR systems has appeared that combine several laser emitters and photodetectors which operate as simultaneous 2-D LiDAR, but with different scanning planes (Fig. 3.2b). These systems allow objects to be scanned similarly to 3-D LiDAR systems. This solution is being widely applied in autonomous vehicles research. Many 3-D LiDAR systems include RGB cameras and provide 3-D point clouds with colour data corresponding to each measured point, in addition to the returning laser beam intensity.

LiDAR sensors estimate the distance to the target by measuring the time between the laser beam emission and the moment at which the light reflected by the target reaches the photodetector, assuming that laser light travels at the constant speed of

Fig. 3.2 The 2-D time-of-flight Hokuyo UTM30-LX-EW LiDAR sensor (**a**) and 3D time-of-flight Velodyne VLP16 LiDAR sensor (**b**) attached to a mast with a GNSS RTK antenna on top. 2D phase shift Hokuyo UTM-04LX 2D LiDAR sensor (**c**)

light through the air (ToF). Usually, the measurement of the elapsed time is done by two different techniques, i.e. time-of-flight and phase shift. A third measuring principle (triangulation) based on the application of trigonometric angular relations to the triangle comprising the emitted and the object reflected laser beams is also used in some LiDAR systems. However, the latter is more suitable for sensors with ranges up to a few metres that is why it is the most used principle in applications with ranges below one metre (Beraldin et al. 2010). Therefore, they are not commonly used in applications that require longer measured distances which is usually the case for agricultural applications.

3.3.3.1   Time-of-Flight LiDAR

To measure the emission–reception elapsed time (from which they give the distance to the target), time-of-flight LiDAR systems emit repetitive pulsed laser beams and measure the time difference between the emission and reception of backscattered light. The electronic measurement of the elapsed time can be done by different methods such as rising edge pulse gradient detection and maximum or constant fraction of peak detection (Beraldin et al. 2010).

An important issue affecting both ToF and phase shift LiDAR sensors is the 'mixed pixels' effect, which appears when the laser beam partially intercepts an object while the remaining beam continues to travel and reaches a second object. In fact, several partial scatterings can occur for a single emitted laser beam, giving rise to several corresponding returning light pulses (also called returns). Some LiDAR systems can discriminate and give the distances for each partially intercepted target or even the full wave returning signal. Other less sophisticated sensors cannot discriminate the returning partial pulses and estimate object distances incorrectly, usually an intermediate value between the partially intercepted objects. This effect occurs mainly at the edges of the measured objects and, thus, it is a common issue when measuring arable crops and trees with LiDAR systems. A consequence of the mixed pixels phenomenon is that the resulting point clouds have points that are aligned forming 'tails' or seem to be 'floating' among the rest of points. To minimize this effect, it is common that software for point cloud processing includes algorithms to filter such points.

The ToF LiDAR sensors usually have long ranges, of around tens or hundreds of metres, and relatively small scanning rates compared to phase shift LiDAR sensors. However, these specifications are evolving quickly with technological improvements and new commercial applications.

3.3.3.2   Phase Shift LiDAR

As an alternative to time-of-flight LiDAR, phase shift or phase measurement LiDAR use continuously emitted laser beams with some of their wave characteristics having been previously modified (Fig. 3.2c). Although several different techniques are used, amplitude modulation (AM) and frequency modulation (FM) are the most popular systems (Behroozpour et al. 2017).

In the case of AM phase shift, the phase displacement is not exclusive for a specific sensor–target distance. In fact, two different sensor–target distances differing by a multiple of the wavelength of the modulated wave will have the same phase shift. Thus, the LiDAR system will not be able to discriminate which is the correct distance and will return the closest to the sensor, i.e. the one shorter than or equal to the emitted wavelength. This phenomenon is known as *range ambiguity* and explains why the maximum measuring range of this type of LiDAR system does not usually exceed more than half the wavelength of the continuous emitted AM laser beam. However, in certain applications some techniques have been developed to solve the ambiguity (disambiguation) and extend the measurement range. According to the inverse relation between wavelength and frequency, the smaller the frequency of the AM laser beam the larger the measurement range of the AM LiDAR system. However, in AM phase shift sensors the measurement error varies inversely to the modulation frequency. Therefore, a balanced modulation frequency should be used for a specific sensor application. Compared to ToF LiDAR sensors, AM phase shift sensors have shorter measurement ranges, with maximum ranges of about 100 m, and much greater scanning rates (Beraldin et al. 2010).

3.3.4 Time-of-Flight Cameras

Cameras based on the time-of-flight principle allow depth information from the captured scene to be determined. A ToF camera is an active sensor consisting of a charge coupling device (CCD) array where the distance from the camera to the object is determined for each of its pixels (Piatti et al. 2012). Such ToF cameras are equipped with several infrared LEDs that emit an amplitude modulated signal (carrier) whose wavelength is usually much longer than the wavelength of the infrared LED. Each pixel from the CCD waffle records the phase of the reflected light and compares it against the emitted one. From the phase difference (as happens with AM LiDAR sensors), each pixel is attached a depth estimate. As expected, the range of ToF cameras is related to the wavelength of the carrier and ranges from 5 to 10 m depending on the manufacturer, with a 'dead-zone' of approximately 0.5 m. The CCD resolution also depends on the model of the camera and the manufacturer. For example, the SwissRanger 4000 has a resolution of 176×144 pixels, and the carrier has a frequency of approximately 20 MHz. The camera can acquire up to 50 FPS (frames per second) and is built for rough environments. Other manufacturers provide ToF cameras with different resolutions, such as Basler (www.baslerweb. com), with a resolution of 640×480 pixels, a range up to 13 m and 20 FPS. On the other hand, LIPS (www.lips-hci.com) has a ToF camera of 8-m range, with a video graphic array (VGA) resolution (640×480 pixels) and 30 FPS. The ToF cameras are designed mainly for indoor use, but most can operate outdoors depending on the lighting conditions. The way ToF cameras work can be summarized as follows:

- All infrared LEDs in the array emit infrared light simultaneously with the same carrier. The emitted pulse takes around 5 to 10 ns.
- Once the carrier is emitted, a time baseline starts running.
- When a pixel receives backscattered light, the phase is measured and subsequent reflections are blocked.
- The system calculates the distances for each pixel in the array according to all the phase differences.

One of the main advantages of ToF cameras is that their hardware does not have any mobile parts and thus they are very robust and able to operate under hard industrial or agricultural conditions. Furthermore, ToF cameras provide very accurate images of the measured scene. The main drawbacks are their short range, their inability to work under direct sunlight and inaccuracy of the depth calculation that may occur when the reflection of the carrier does not come directly from the target (Erz and Jähne 2009).

Recently, some devices have merged the principles of ToF and RGB cameras in a single device, such as the Microsoft Kinect version 2 (Fig. 3.3). This device consists of a passive RGB camera and an active ToF camera. As a result, the sensor output is a point cloud with up to seven variables associated to each of the points or pixels: (1) location information, usually x, y and z local Cartesian coordinates related to the sensor origin of coordinates, (2) colour information, usually RGB and (3) infrared

Fig. 3.3 The Microsoft Kinect version 2 is a low-cost consumer grade RGB-depth camera (ToF and RGB camera) with potential applications in agriculture. Microsoft Kinect version 2 acquiring data in an apple orchard (*Malus domestica* Borkh.) (**a** and **b**), 3-D point cloud displaying RGB colour data (blue points at the bottom area are lacking RGB data because of the different fields of view of the RGB and IR sensor (**c**) and 3-D point cloud displaying infrared returning intensity to the sensor (**d**)

intensity of the backscattered light (IR). Thus, the additional colour and IR data in the point clouds provide more information to enable the characteristics of the target to be extracted by post-processing algorithms. Sometimes cameras that provide RGB data and distance are also called depth cameras or RGB-D cameras because they estimate the depth of each pixel in the recorded scene.

3.3.5 Structured Light Sensors

Structured light sensors are another type of depth camera providing RGB and distance (or depth) data on a per pixel basis. These kinds of sensors are active and comprise a Near Infrared (NIR) laser source emitting a predetermined structured light pattern, an NIR digital camera, an RGB digital camera and a microprocessor programmed with machine vision processing algorithms for pattern recognition.

The output of the sensor is a point cloud with x, y, z Cartesian coordinates for each point referenced to a coordinates origin placed in the sensor's focal point. In addition, each point has its RGB colour values, obtained from the embedded RGB digital camera (Fig. 3.4). Some limitations arise from the structured light working principle. In highly illuminated environments, especially with background NIR light as for sunlight, the IR sensor camera loses accuracy in detecting the NIR pattern, hence depth measurements are inaccurate (Andújar et al. 2016; Rosell-Polo et al. 2015). This drawback greatly limits the use of these sensors outdoors, especially on sunny days, and restricts their use to before sunrise, after sunset or at night (in that case an artificial light source should be used to obtain the RGB data of the acquired points, if required). Another common limitation of structured light sensors concerns their ability to detect thin or slim objects. This depends on the distance to the object and it is caused by the need to have a minimum area of light pattern for the algorithms to estimate the distance properly. An advantage of structured light sensors compared to LiDAR sensors and ToF cameras is that they usually do not suffer from the mixed

Fig. 3.4 Microsoft Kinect version 1, a low-cost consumer grade depth camera (structured light sensor and RGB camera) with potential applications in agriculture (**a**) and different views and details of 3-D point cloud with RGB data for each pixel of a medlar tree (*Mespilus germanica* L.) captured by the Kinect v1 depth camera (**b, c** and **d**)

pixels phenomenon and, thus, much *cleaner* point clouds are obtained because of the absence of the noise associated to this effect.

3.4   Data Processing

The second stage consists of preparing and processing the acquired data from the sensors to extract useful information for advisors and farmers to make more informed agronomic decisions. Such information may be displayed as maps to be interpreted in the so called map-based PA approach or may be used on a real-time basis (real-time sensor-based approach). The former requires more knowledge and training in several computer programs such as GIS, but allows more complex decisions to be made and local agronomic knowledge to be included in the decision-making process. The latter uses simpler embedded solutions taken in milliseconds. Robots can be used in PA in both approaches. Service units could be used for scouting (map-based PA) or to perform agricultural tasks (real-time sensor-based PA).

Sensors are mainly used for monitoring and inspection of the farm (scouting), therefore, it becomes essential to know beforehand which variable or set of variables to observe. Once the variables of interest are established, the next step is to choose the most appropriate sensor to fulfil the monitoring or inspection task and to process the information provided by them.

The way data are processed might vary according to the application and the hardware being used. At first, we can distinguish clearly between two different data processing paradigms: real-time and batch data processing. Most algorithms are designed as Finite State Machines, therefore, time becomes the core of each approach. However, not all sensors and algorithms operate simultaneously. Delays, time propagation between stages and glitches often occur and have to be taken into account when processing data. In fact, most stages in every data processing implementation work asynchronously, which represents an important challenge to be faced by the designer or the researcher. Batch processing is used in map-based PA, where time elapsed between data acquisition, analysis, decision making and operation is not critical. Real-time processing is used in sensor-based real-time PA where time between sensing and operation takes only several milliseconds.

3.4.1   Point Cloud Creation

Most of the sensing principles and instruments used in crop and environmental modelling presented above provide 3-D spatial coordinates of measured points of the target either in Cartesian or rectangular coordinates (x,y,z) or in polar, spherical or cylindrical coordinates (polar distances and angles). Usually, this 3-D information is referred to an origin of coordinates located in the sensor itself. By representing the 3-D coordinates graphically of each measured point of the object or scene a so-called point cloud is obtained. Point clouds are, therefore, the most common output of the measurements undertaken with sensors and systems with 3-D measuring capabilities. For this reason, Sects. 3.4.1 and 3.4.2 will describe the creation and processing of point clouds.

Each sensor has its own reference system with a local origin of coordinates, usually within the sensing device. When scanning with a 3-D system from a single stationary position a local reference system may be sufficient, depending on the application. However, other research, including agricultural applications, is based on the movement of 2-D or 3-D sensing systems throughout the area of interest to obtain point clouds in absolute terrestrial coordinates, usually in the UTM (*Universal Transverse Mercator*) system for further analyses. In the next sections, several commonly used approaches to create georeferenced point clouds are introduced.

3.4.1.1 Absolute Geolocation

One of the procedures used most to obtain point clouds in absolute coordinates uses global navigation satellite systems (GNSS) such as GPS, GLONASS, GALILEO and BEIDOU. The GNSS systems provide the coordinates of the receiving antenna and are referred to an absolute terrestrial coordinate system, either in terms of geographical or projected coordinates. In this way, if the position of the local origin of coordinates of the 3-D measuring system relative to the GNSS antenna is known and remains invariable while the measurements are carried out, it is possible to obtain the coordinates of each point in absolute terrestrial coordinates. Measurement errors of millimetres or a few centimetres may be achieved with 3-D sensing systems, therefore, an important requirement with the GNSS approach is that errors should also be limited to a few millimetres or centimetres to obtain reliable absolute georeferenced point clouds. This leads to the use of GNSS systems with enhanced accuracy derived from ground-based augmentation systems, such as Real Time Kinematic (RTK). According to manufacturers, such receivers may give horizontal errors of around 1 to 6 cm and vertical errors from 3 to 8 cm depending on the specific solution and operating conditions. In most agricultural applications, this level of accuracy may be adequate, but when greater accuracy is required (accuracy of few millimetres) other location solutions should be used (e.g. surveying techniques).

Measurements in the field done by 3-D sensing systems may be affected by other sources of error such as system vibration, misalignments and direction changes, among others. Several procedures are available to correct these issues, at least partially. Figure 3.5a shows a mobile terrestrial laser scanner on an electric vehicle designed for autonomous crop scanning as a scouting robot. Dynamic misalignments can be attenuated by mounting the 2-D or 3-D sensor on a gimbal type mechanism to keep the sensor level regardless of changes in the slope causing either pitch or roll (Fig. 3.5b and c). In addition, gimbal systems compensate considerably for inaccuracies produced by system vibrations. On the other hand, changes in direction can be approached by continuously correcting the direction of movement by calculating the

Fig. 3.5 Mobile terrestrial laser scanner mounted on an electric vehicle designed for autonomous crop monitoring (**a**) based on a 2-D LiDAR sensor mounted on a gimbal to keep it level (**b** and **c**) and based on an RTK GNSS receiver to georeference the sensor position. Prototype developed by the Research Group on AgroICT & Precision Agriculture from the Universitat de Lleida-Agrotecnio-CERCA Centre

instant displacement direction from the measured GNSS absolute terrestrial coordinates in the actual system position and those corresponding to the system location at a previous instant. This approach can provide acceptable corrected point cloud coordinates provided the direction of the system movement does not experience considerable or abrupt changes, in which case values of the real point coordinates would appear distorted. Another approach to improve the point cloud accuracy, not only with regard to changes in direction but also to correct system vibration, is based on the use of inertial measurement units (IMU) attached to the sensor. The IMUs mainly comprise accelerometers, gyroscopes (both in two or three dimensions) and sometimes magnetometers and barometers. An IMU becomes useful to estimate motion because information provided by the sensors enables the translational and rotational components to be found of the moving rigid body at which the IMU is attached. The latter is called inertial navigation (Mutz et al. 2016; Siegwart et al. 2011). If at a specific time instant the GNSS absolute coordinates and the real time orientation of the 3-D sensor are known, proper transformation equations and algorithms can be applied to calculate the absolute coordinates of each measured point. The IMU + GNSS-based techniques are probably the most accurate and smartest way to obtain absolute georeferenced point clouds. Two requirements are needed for this. On the one hand, the timestamps of GNSS, IMU and 3-D sensing systems should be synchronized precisely, but on the other hand, strong accuracy is required for the IMU sensor as angular direction errors greatly affect the accuracy of the point coordinates. The latter leads to a marked increase in the cost of accurate IMU-based systems.

3.4.1.2 Simultaneous Location and Mapping

The simultaneous location and mapping (SLAM) algorithm is mainly used in real-time data path processing approaches. However, when the focus is on the mapping process and the localization problem can be solved for short-time navigation horizons, then SLAM can also be used with batch processing data paths. The need for a short-time navigation system, such as an RTK, dead-reckoning or an IMU-based system (for inertial navigation purposes) exists when the sensor and or the service unit is moving or interacting with the environment during data acquisition. Once acquired, data are processed as a batch which may require substantial time to extract features. Thus the need for a location system with local consistency boundaries (i.e. it works for short-time positioning purposes). Once data have been collected, the SLAM algorithm performs as shown in Fig. 3.6 (real-time SLAM).

The problem with SLAM for batch processing is depicted in Fig. 3.7. The red triangles represent the data acquisition points and the dashed black line is the path followed by the sensor or the service unit along an alley in a grove. The solid blue lines represent the distances travelled by the sensor or the service unit during which the system processes data and does not acquire new information. Such distances are not necessarily the same since the processing time is strongly correlated to the amount of data being processed. When stochastic approaches such as the Extended Kalman Filter, Particle Filters or the Information Filter (among others) are used and

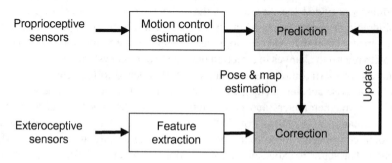

Fig. 3.6 Layout of the real-time SLAM algorithm

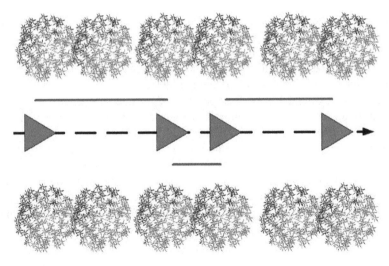

Fig. 3.7 Representation of the problem of SLAM using batch data: the system remains in an open loop during the information processing (solid blue line). The red triangles represent the data acquisition points, the dashed black line is the path followed by the sensor or the service unit

the layout of the SLAM algorithm shown in Fig. 3.6 is taken into account, what actually happens inside the algorithm is that it remains in the prediction stage using the proprioceptive information until enough data have been collected to be able to execute a batch processing algorithm. When the outcome is available, then the SLAM algorithm performs a correction stage. As in the real-time case, the SLAM algorithm always delivers the 'pose' (position and orientation of the sensor or the service unit), the map of the surrounding environment and the corresponding covariance matrices that link the map with the proprioceptive information managed by the system.

The main benefit of using a batch processing strategy with a SLAM algorithm instead of a real-time processing data path is that more complicated features from the environment can be acquired and extracted. An example of this type of SLAM is shown in Cheein et al. (2011) where a grove is scanned using a 3-D LiDAR-based system on a service unit. The data are processed and the canopy volume of each tree crown is estimated as well as the position of the service unit at each acquisition point.

3.4.1.3 Stereophotogrammetry

Another technique used to create point clouds is stereophotogrammetry. According to the International Society for Photogrammetry and Remote Sensing, "photogrammetry is the art, science, and technology of obtaining reliable information from noncontact imaging systems of the Earth and physical objects and processes through recording, measuring, analysing and representation". Stereophotogrammetry is a variant that enables the 3-D coordinates of pixels to be estimated from two or more

RGB or multispectral images of a scene taken from different points of view. Specific programs are used to batch process sets of pictures. The first step is to identify common points shared by two or more pictures. In a second step, a mosaic is created for the whole scene. In a third step, triangulation is used to determine the coordinates of each single pixel. Pixel location data are added to already existing RGB data. The output of the process is a colour point cloud of the scene in local coordinates. When absolute coordinates are required, previously georeferenced targets should be distributed throughout the scene to be captured with the camera and their absolute coordinates should be entered into the process program. Once the point cloud of the agricultural scene has been created, analytical tools will be applied to extract geometric, structural and other georeferenced information of the crop and its environment as done with point clouds created by different techniques. In agriculture, stereophotogrammetry can be used with images taken from unmanned aerial vehicles (UAV) or from terrestrial vehicles.

3.4.2 Point Cloud Analysis

Point clouds are the basic information derived from the previously described sensor data and techniques. As explained above, data from 3-D sensing systems can be used both for navigation and for acquiring, characterizing and classifying agricultural scenes. All these purposes can be addressed considering, at least, two variables: time and space. On the one hand, with regard to the time scale, two main approaches can usually be considered: A) real-time applications and B) batch processing applications. On the other hand, with regard to spatial scale there are four possible approaches: (1) point-by-point analysis, i.e. the processing, analysis and decision making are made for each individual detected point, (2) scan-by-scan analysis, i.e. the processing, analysis and decision-making are made for each scan (a scan comprises all the detected points in a single LiDAR reading in a plane, vertically, horizontally or obliquely oriented; this approach is only feasible when using 2-D LiDAR sensors or when 3-D sensors can differentiate 2-D scanning planes), (3) partial or space-limited point cloud analysis, which consists of the analysis of a spatially-limited point cloud or simply a subset of the whole scene's point cloud and finally (4) analysis of the point cloud of an entire tree row or even a crop plot as a whole. While some approaches based on time and spatial scales can be easily combined, some others are difficult to merge. For example, approaches (A) and (1) leading to real-time point-by-point applications, as well as approaches (A) and (2) real-time scan-by-scan applications are feasible. However, it is unrealistic that approaches (A) and (4), i.e. real-time whole point cloud analysis, could be simultaneously affordable in the short term because, at present, the analysis of the point cloud as a whole needs lengthy post-processing times, especially when large areas have been scouted. Although each feasible combination of time and spatial approaches has its own specific processing methods and algorithms, in this section we discuss some considerations regarding the point cloud analysis that will be partially or totally applicable to approaches (2), (3) and (4).

Although point clouds can be processed manually, it is more usual and convenient to address this stage through computer-based automated procedures. Two methods will be presented: *a deterministic-based approach* and *a stochastic-based approach*. Two alternative approaches can be followed with computer-based point cloud analysis: (1) using dedicated software developed by the user; an advantage of this approach is that the software can be designed to meet the requirements of each specific application and, therefore, will be very efficient; however, this approach requires resources and computer programming abilities. The second approach consists of (2) using commercially available software specifically designed for point cloud processing; an advantage of this option is that these processing tools are ready to be used, which will save time once the user has learned how to use them. The counterpart is that many programs are not free to use although some are. Another issue is that many point cloud file formats are available that depend on the kind of data stored for each measured point (for instance, some formats include only x,y,z point data, while others add much more information such as R,G,B, IR intensity, normal components of x, y and z, among others). Not all point cloud formats are supported by all commercial software. Some are more extended while others have a more restrictive use. Although the ASPRS las and laz formats are becoming a standard, many other point cloud files formats are identified by different file extensions, such as txt, asc, pcd, neu, xyz, pts, csv, e57, ptx, vtk, pn, pv, pov, icm, dp, rdb, rds and others.

3.4.2.1   Deterministic Approach

In the deterministic approach, the point cloud analysis relies on the absence of random variables or random interactions, in such a way that a deterministic algorithm will lead to exactly the same output results from a given point cloud (regardless of its initial state or starting variable).

Although not imperative, most deterministic approach algorithms assume that the spatial coordinates of each single point obtained by the 3-D sensing system, either local or absolute, are known and thus treated as initial data. The algorithms and computational steps to extract the point cloud variables of interest are usually based upon a combination of mathematical expressions, frequently from the field of geometry, and common programming structures and procedures. Three types of information are of interest in agricultural applications: (1) the geometry of crops and trees, (2) their structure and (3) their components. These three levels of information are used, although with different degrees of development, by agricultural scientists and technical staff to advance towards a more economically and environmentally sustainable management of crops. The geometric properties of interest include, among others, crop height, width, canopy volume in general, and, specifically in tree crops, leaf wall area, tree volume and projected tree horizontal area (Escolà et al. 2016). On the other hand, the structural features can include light penetrability, leafiness, porosity (Escolà et al. 2016) and biomass (Andújar et al. 2016). Finally, characterization of the components of plants and crops involves the detection and counting of fruits and flowers (for thinning and prediction of yield estimates), the characterization of trunks and branches to optimize tree training and crop management (Méndez et al. 2016), and the detection and mapping of weeds (Andújar et al. 2013), among others. This last issue is known as *classification* of point cloud components. Moreover, the above properties can be computed at different spatial scales, including the whole plot, a single row, a single tree or plant and a portion of a plant (for example, at different heights and widths).

3.4.2.2   Stochastic Approach

Iterative closest point (ICP) algorithms treat the information in a deterministic way; therefore, the outcome will be the same for the same two point clouds under study. However, when adding the covariance of the error associated with the sensor measurements, then it is unlikely to have two equal readings for the same scenario. In such situations, when the sensor readings and point cloud are random variables, other ICP approaches are used to include probability and covariance (or volume of uncertainty) as possible metrics. Such is the case of the Sum of Gaussian Scan Correlation, which is an ICP algorithm that produces homogeneous transformation matrices when the strongest correlation among point clouds is achieved. The Probabilistic Iterative Correspondence uses probability as a metric between point clouds and Least Squares based approaches that include the covariance information. For further discussion, the reader is encouraged to read Rusinkiewicz and Levoy (2001).

3.5   Applications

This section presents relevant illustrative examples and applications of agricultural robotic sensors for crop and environmental modelling. This application list is not intended to be exhaustive in the sense that many other applications have already been developed or are being implemented by many research groups and companies. The selected applications have been grouped into four subjects, according to their final objective: i) service unit navigation, ii) terrain modelling, iii) canopy characterization and mapping, and iv) weed detection and mapping.

3.5.1   Service Unit Navigation

A service unit (i.e. an automated piece of machinery for industrial or production processes) can be used to perform many tasks autonomously related to monitoring, supervision and inspection of agricultural scenarios, as stated in Cheein and Carelli (2013). With the new era of information and communication technologies, robots have become the carriers of new sensors to replace the manual data acquisition process. A robotic service unit can work day and night, whereas for humans energy

Fig. 3.8 Service units in agricultural environments: an all-terrain vehicle used for agricultural purposes in Chile (**a**); a Pioneer 3AT robot (**b**) specially equipped for scouting in an avocado grove in Chile (**c**)

consumption is the main constraint. One of the key achievements is the ability to georeference data automatically with GPS or GNSS receivers, or any other positioning system. Thus, it is possible to track agricultural information in space and in time. Figure 3.8a shows an automated all-terrain vehicle adapted to work in agricultural environments. The robotization of the vehicle was performed by the Pontificia Universidad Católica, Chile, (Aguilera-Marinovic et al. 2016). Such a vehicle can move autonomously along alleys in a grove and is equipped with different types of sensors, a differential GNSS receiver and several IMUs strategically located on the chassis. On the other hand, the Pioneer 3AT shown in Fig. 3.8b is an example of how to use robotic platforms to acquire and map information in an avocado (*Persea Americana* Mill.) grove in Chile with a Hokuyo UTM-30LX LiDAR sensor, a commercial GPS receiver with accuracy of two metres and a laptop computer for data processing (SLAM) and motion control. In addition, the Pioneer 3AT has a Microsoft Kinect version 2 (Kinect v2) sensor at the front for terrain characterization.

When controlling the motion of a robotic service unit, terrain characteristics and environmental constraints (such as location of trees and field workers moving around the vehicle) are important factors to be taken into account. For example, previous work by Cheein (2016) showed how to design and implement motion controllers in service units to drive autonomously within groves to acquire important information for the farmer. An example of a path followed by a robot using the motion controllers published in Cheein and Scaglia (2014) and the Pioneer 3AT as a service unit is shown in Fig. 3.8c, where the platform had to traverse around an avocado grove. More information about service units and their corresponding design specifications can be found in Cheein and Carelli (2013) and in its cited references.

3.5.2   Terrain and Soil Modelling

Terrain, regarded as the top surface of the soil, is important to both robot traversability and agricultural uses. Sensing the terrain can be done directly or indirectly with the sensors described in previous sections. Direct techniques are used when the terrain is the target of the sensing operation, whereas indirect techniques obtain measurements from the terrain as a secondary product. Digital terrain models provide basic information in map-based PA because topography or relief can explain some of the crop variation when it comes to soil characteristics or water and nutrient movements within the field from differences in elevation.

3.5.2.1   Terrain and Soil Characterization

When dealing with service unit traversability, knowing the terrain characteristics provides many advantages that have a direct effect on productivity. For example, by including information of the terrain characteristics, a service unit can reduce energy consumption by up to 5 %, as shown in Cheein et al. (2015). The topology and morphology of the terrain can be interpreted by the robotic sensors to classify the nature of the terrain. Traversing over sand or over gravel require different ways to move and energy consumption will vary. To overcome this issue, both exteroceptive and proprioceptive sensors might provide enough information to characterize the terrain. For example, Yandun et al. (2016) used a Kinect v2 outdoors for real-time terrain classification. In that research, the system could distinguish and classify up to six different terrains found in an agricultural environment with up to 95 % success using RGB, depth and intensity data. That information was then used to self-tune the motion controller properties of a service unit to save energy and to avoid slippage and sinking situations. For example, Fig. 3.9 shows six snapshots of different terrains

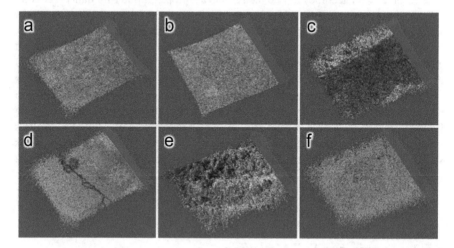

Fig. 3.9 Six snapshots of terrain classification using the Kinect v2. During the trials, six different terrains were detected: sandy (**a**), gravel and weeds (**b**), sole weeds (**c**), sole mud (**d**), mud and gravel (**e**), and clay (**f**). Adapted from Yandun et al. (2016)

detected and classified using a Kinect v2 mounted on the Pioneer 3AT unit while navigating (Fig. 3.8b).

Furthermore, Marinello et al. (2015) used a structured light camera (Microsoft Kinect v1) to assess the soil surface roughness and related it to different tillage operations and intensities. Their system acquired depth data on-the-go from the Kinect sensor to be analysed subsequently and extracted objective and repeatable data to characterize the soil surface.

To characterize the soil in depth, service units can be used to collect georeferenced soil samples autonomously, which requires robust and powerful actuators, or indirect soil sampling methods can be used to derive maps of soil variation. The latter can take advantage of geophysics sensing principles such as electromagnetic induction. This kind of sensor does not require direct contact with the soil and hence the service unit has only to pull a sled with a load of a few kilograms. The sensor estimates soil electrical conductivity on-the-go. This property is related to many others such as soil moisture, structure, compaction, clay content, salinity and some others. While it is difficult to classify the soil from raw electrical conductivity values, they do provide information on whether the soil in the field or orchard is uniform or variable from the mapped values. It is then up to the farmer or advisor to opt for a uniform management approach or for variable-rate solutions.

3.5.3 Canopy Characterization and Mapping

Canopy characterization and subsequent crop and environmental modelling is important to improve agricultural management through increasing efficiency of agricultural operations. Obtaining accurate key geometric and structural characteristics of crops is a giant step forward, which makes it possible to use such information for decision making in crop management. On the other hand, most of the sensors described above are used in *proximal sensing*, although some can also be adapted for use in *remote sensing*. Indeed canopy and crop characterization with proximal sensors, despite being more labour and time consuming than remote sensors, can supply much more detailed and accurate information, with much finer spatial resolution. Examples and applications of ground-based canopy characterization and mapping for each sensing technology are presented in the following sections. The systems described in this section are based on exteroceptive sensors, which means that they could be used by robots but also by other non-robotic equipment for crop and environmental modelling. When the sensor is part of a robot, the data acquired could also be used for robot navigation. Crop and environmental modelling is of great interest in precision agriculture, but it is rarely implemented in commercial robots. The robotization of such technologies will take crop scouting to a higher level allowing robots to be sent for periodic crop monitoring of the whole orchard or, alternatively, to revisit specific locations for high resolution and more accurate measurements. In this section we list some of the sensing techniques already in use but also some potential sensors to be implemented in robots to stimulate its implementation.

3.5.3.1 Ultrasonic Sensor Techniques

In tree crops, ultrasonic sensors (US) were used in the early 1980s to characterize orchard canopies electronically to estimate the so-called tree-row-volume to adjust pesticide dose rates. In 1988, a commercially available orchard sprayer (Roper Grower's Cooperative, Winter Garden, FL, USA) had five ultrasonic sensors mounted on each side of the sprayer to detect the presence or absence of vegetation at different heights in real time to turn spraying on and off. Giles et al. (1988) developed an electronic system with three ultrasonic sensors on each side of a sprayer to estimate the tree-row-volume and reported success in field tests. Later, Schumann and Zaman (2005) used a commercial Durand–Wayland (Durand-Wayland, Inc., Lagrange, Ga, USA) ultrasonic system consisting of 10 sensors at different heights ranging from 0.6 to 6.0 m to measure and map canopy volume in citrus groves for application in PA. In 2011, Escolà et al. conducted field tests to validate the accuracy and robustness of ultrasonic sensors under commercial orchard conditions. The specifications of each sensor might affect its ability to detect the canopy. For example, the sound cone footprint size will determine the size of the detectable canopy gaps. As the sound cone footprint is directly related to the sound cone aperture (sensor specification) and the distance to the target (operation of the sensor), it is important to choose the proper sensor correctly and to use it in the most effective way. In addition, it is important to choose the most suitable number of sensors and place them at the most appropriate height to estimate the canopy cross sections accurately (Fig. 3.10).

In addition to the estimation of canopy volume, Balsari et al. (2008) and other authors developed systems to estimate the thickness or density of the canopy. These systems analysed the amplitude or intensity of the waves returning to the sensor. An attenuated echo would come from a low density canopy, whereas high-amplitude and powerful signals would come from denser canopies.

In arable crops, height measurement using ultrasonic sensors is more common. Many mobile systems have been developed for on-the-go measurements. Research has been published with indirect estimates derived from crop height readings. Maertens et al. (2003) developed a system for crop density estimates from the returning ultrasonic echoes. Other authors use US sensors to estimate tiller density

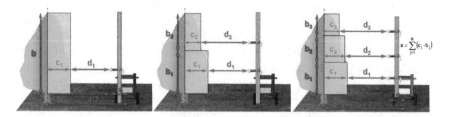

Fig. 3.10 Canopy semi-cross section estimated by 1, 2 or 3 ultrasonic sensors assuming the sensors are located at a distance e from the centreline of the alleyway. With more sensors, vertical resolution is improved and the cross-sectional area is estimated more accurately according to the formula provided

and leaf area index in winter wheat (Scotford and Miller 2004), whereas many other studies correlate crop height with biomass (e.g. Reusch 2009).

3.5.3.2   LiDAR Sensing Techniques

In tree crops, as explained in Sect. 3.3, canopy characterization with LiDAR sensors was initially associated with the optimization of spray application techniques. Soon, measurement of the geometric and structural properties of plants obtained by terrestrial laser scanners or by mobile terrestrial laser scanners to improve agricultural management became a main objective per se. Pioneer research focused on system development (Tumbo et al. 2002; Walklate 1989; Wei and Salyani 2004) and on making first attempts to measure geometric (height, width, volume) and foliar density of isolated or short tree sections (Palacín et al. 2007; Wei and Salyani 2005), whereas Llorens et al. (2011), Rosell et al. (2009) and Sanz-Cortiella et al. (2011) were among the first to obtain 3-D point clouds of entire rows and plots in fruit orchards. In their early research, these authors placed known geometric objects in the scanned scenes which were used as reference objects for merging the point clouds obtained from both sides of a row. Subsequently, reference objects were no longer necessary with the introduction of a RTK–GNSS receiver in the experimental system. This enabled UTM-coordinate georeferenced point clouds to be obtained with an accuracy of few centimetres. Much research about canopy characterization of different fruit species with LiDAR scanning has been published recently, showing the important role of LiDAR sensors in canopy characterization and mapping (del-Moral-Martínez et al. 2016; Underwood et al. 2016). In Escolà et al. (2016), in addition to determining canopy height, width, volume and porosity every 10 cm along the rows, the authors displayed the results using raster maps for farmers and advisors so that they could be easily interpreted. Moreover, as they created canopy volume maps on different dates, growth maps were obtained by subtraction of raster information (Fig. 3.11).

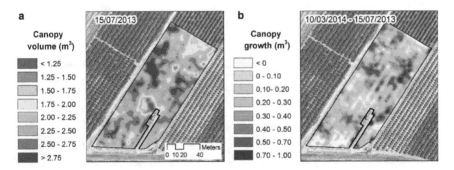

Fig. 3.11 Canopy volume map of a 1-ha intensive olive orchard derived from a 2-D LiDAR-based MTLS obtained at a single date (**a**). When a map of a different date is available, subtraction of the raster maps produces a canopy growth map (**b**) between the two dates (adapted from Escolà et al. 2016)

With such information, farmers would know what areas in his or her orchard produce larger or smaller trees in an objective and quantifiable way. Once growth maps are created, it is time to go to the orchard and to try to understand the reasons for such behaviour and to make management decisions accordingly. A similar system to that used by Escolà et al. (2016) to scan the orchard was used in Méndez et al. (2016) to analyse the woody structure of deciduous fruit trees trained in different patterns. The number of branches and their length were computed before and after pruning and the results were correlated to pruning weight. Such information could be used by farmers to correlate pruning intensity with yield and to decide on pruning intensity, which is now considered challenging, time consuming and inaccurate when performed manually.

In arable crops, other authors have used LiDAR sensors to estimate crop biomass for different crops (Ehlert et al. 2008). Saeys et al. (2009) developed a system to estimate crop density and volume in real time at the entrance of combine harvesters. Gebbers et al. (2011) proved that LiDAR sensors could be used to estimate leaf area index in winter wheat.

3.5.3.3 ToF Cameras Techniques

Point clouds of plants and crops obtained with ToF cameras alone are usually characterized by large amounts of noise which make it difficult to extract useful information from an agricultural management perspective. Figure 3.12 shows front and side views of the point cloud of a medlar tree (*Mespilus germanica* L.) and several bushes obtained with a PMD[vision] CamCube 2.0 ToF camera with 204×204 pixel resolution, $60° \times 60°$ field of vision and 0.3 to 7 m range. From a practical point of view, successful canopy characterization and mapping with ToF cameras was delayed until the emergence of ToF–RGB systems, such as the Microsoft Kinect version 2 sensor, launched in 2013. As previously explained, the latter is a mass production inexpensive ToF camera, which includes RGB and IR data with greater pixel resolution (as large as 512×424), larger field of view ($70° \times 60°$), similar range and usually smaller noise levels than the best existing ToF cameras at the moment (2013). Although

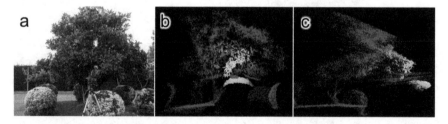

Fig. 3.12 Picture of the data acquisition process with the ToF camera in the foreground on a tripod (**a**). Front (**b**) and side (**c**) views of the point cloud of a medlar tree and some quasi-spherical shaped bushes obtained with a PMD[vision] CamCube 2.0 ToF camera. Colours correspond to backscattered light intensity (larger in red)

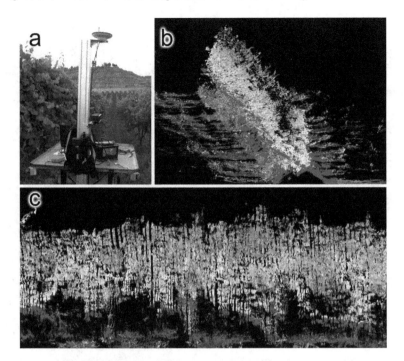

Fig. 3.13 Microsoft Kinect v2 sensor undertaking measurements in a vineyard (**a**), perspective view of the RGB point cloud obtained (**b**) and side view of the RGB point cloud (**c**). Blue coloured points have XYZ data but no RGB data because of the different field of view of the RGB and IR cameras of the Kinect v2 sensor

both sensors experience the mixed pixels phenomenon and ToF cameras usually have better tolerance to large outdoor illuminance values, the remaining outstanding properties of Microsoft Kinect v2 sensor and similar sensors using the same principle have almost displaced the use of ToF cameras in canopy characterization and mapping. Figure 3.13 illustrates the use of Microsoft Kinect v2 sensor undertaking measurements in a vineyard as well as a sample of the point cloud obtained.

3.5.3.4 Structured Light Sensor Techniques

Although several research projects and applications concerning plant characterization with structured light sensors have been published since these sensors were available (Chéné et al. 2012; Yamamoto et al. 2012), few measurements have focused on real canopy characterization and mapping measurements. This is mainly because these sensors fail to perform reliable measurements outdoors in daylight (Azzari et al. 2013; Rosell-Polo et al. 2015; Wang and Zhang 2013). Rosell-Polo et al. (2015) studied the potential of Microsoft Kinect version 1 in agricultural applications and quantified the effect of natural light when used outdoors. To illustrate the result of

Fig. 3.14 Row sections of an apple orchard (**a**) and point cloud obtained with a Microsoft Kinect version 1 sensor (**b**)

canopy characterization with structured light sensors, Fig. 3.14 shows a point cloud of a section of an apple (*Malus domestica* Borkh.) tree row obtained at night with artificial lighting.

3.5.3.5 RGB Cameras and Stereophotogrammetry Techniques

The RGB cameras and stereophotogrammetry are widely used with UAVs. However, in intensive tree plantations nadir photographs from UAVs might not represent correctly the bottom sides of tall tree rows. Thus, as RGB cameras are relatively cheap compared to other sensors, and software tools are available for stereophotogrammetry processing, attempts have been made to implement sideward canopy orchard characterization. Nevertheless, the short distance between rows in intensive fruit orchards makes it necessary to acquire several images per linear metre of row and possibly at different heights to achieve the required swath to allow depth to be computed. When this is feasible, the results are impressive because the derived point clouds are dense and clean (without the mixed pixels effects of some LiDAR sensors) and, in addition to 3-D coordinates, RGB data are also included on a per point basis.

3.5.4 Weed Detection and Mapping

Weed detection and mapping is an important operation in PA of arable crops. Weeds often follow spatially- and temporally-dependent patterns, therefore it is worthwhile to have an idea of what their spatial distribution is so that variable-rate herbicide application could be considered. Some solutions are not map-based but on-the-go. That is, the system detects and acts on the weeds in real-time to get rid of them instantly by using different strategies (herbicide spraying, mechanical removal or physical destruction). Examples and applications of weed detection and mapping solutions are presented in this section.

Ultrasonic sensing for weed detection between row crops is a cheap and reliable solution for robotic scouting in arable crops. Andújar et al. (2011) tested a proof of concept with nadir ultrasonic height measurements in the inter-rows of a maize field. Anything significantly higher than bare soil was considered a weed. There were marked differences between grass-like and broad-leaved weeds. When measurements were made at early stages, maize leaves did not interfere with weed detection, but difficulties occurred when trying to discriminate young weeds from bare soil.

Several systems have been developed to sense and classify weeds using RGB cameras mounted on robotic platforms and machine vision processing. Some of them are used for weed scouting and mapping and many others act on them by spraying herbicide (Berge et al. 2012), destroying them (Blasco et al. 2002, used an electrical discharge) or mechanically removing them. Slaughter et al. (2008) provided a review of autonomous robotic weeding systems. In Chap. 7, a cooperative fleet of robots use RGB cameras on board UAVs and tractors to detect and subsequently remove weeds.

Commercial systems are already available that use radiometric multispectral sensors, to detect and spray weeds. These sensors could be mounted on service units to scout the fields to create weed infestation maps. Some of these sensors are WeedSeeker (Trimble) and WEEDit, which are widespread in PA weeding operations.

The ability of mobile terrestrial LiDAR sensors to discriminate weeds from crops and soil and its potential to characterize weed vegetation has also been demonstrated (e.g. Andújar et al. 2013). Tamás et al. (2014) used a 3-D airborne laser scanner combined with a hyperspectral imaging system and aerial photographs to measure and investigate weed cover. They could detect emerging weeds and found that airborne LiDAR was good for detecting weeds, but could not identify weed species. The authors concluded that LiDAR scanning is an effective technique to scout and characterize weeds; terrestrial LiDAR systems are more suitable for classifying weed species than airborne systems because of the higher point density of the former.

The ToF cameras have also been used to detect and discriminate weeds, especially since the release of Microsoft Kinect v2. Many studies have been undertaken to assess the ability and potential of ToF cameras, alone or in combination with other sensing systems, for scouting, quantifying, discriminating and mapping weeds, and more generally, to characterize plants (Hansen et al. 2013; Ruckelshausen et al. 2009). The ability of Microsoft Kinect v2 sensor, and other similar recent sensors, to add colour and infrared information to point clouds using a single sensor is a powerful advantage which can potentially facilitate the crop and weed classification and discrimination. Gai et al. (2015) studied the plant recognition potential of the Microsoft Kinect v2 sensor for weed management by fusing 2-D textural information from the RGB Kinect camera and the 3-D depth information obtained from point cloud data. The authors concluded that this sensor was promising and reliable for sensing the environment by agricultural robots, as well as for discriminating and locating plants. In a similar way, Andújar et al. (2016) studied the potential benefits of the Microsoft Kinect v2 sensor to identify and discriminate weeds from a crop in weed-infested maize crops under real field conditions. The colour images supplied by the embedded RGB camera allowed short weeds to be discriminated from the soil surface. Estimated volumes of maize and weeds showed good correlations with manually measured maize and weed biomass.

Acknowledgements The authors thank the University of Lleida (Catalonia, Spain) and Universidad Técnica Federico Santa María (Valparaíso, Chile) for their support and the following projects and grants which funded the authors' research: projects AGL2002-04260-C04-02, AGL2007-66093-C04-03, AGL2010-22304-04-C03-03, and AGL2013-48297-C2-2-R from the Spanish Ministry of Economy and Competitiveness and EU FEDER, and projects Fondecyt Regular 1171431 and CONICYT FB0008 from Chile.

References

Aguilera-Marinovic S, Torres-Torriti M, Cheein FAA (2016) General dynamic model for skid-steer mobile manipulators with wheel-ground interactions. IEEE/ASME Trans Mechatron 22:1–1. https://doi.org/10.1109/TMECH.2016.2601308

Andújar D, Escolà A, Dorado J, Fernández-Quintanilla C (2011) Weed discrimination using ultrasonic sensors. Weed Res 51:543–547. https://doi.org/10.1111/j.1365-3180.2011.00876.x

Andújar D, Escolà A, Rosell-Polo JR, Sanz R, Rueda-Ayala V, Fernández-Quintanilla C, Ribeiro A, Dorado J (2016a) A LiDAR-based system to assess poplar biomass. Gesunde Pflanzen 68:155–162. https://doi.org/10.1007/s10343-016-0369-1

Andújar D, Ribeiro A, Fernández-Quintanilla C, Dorado J (2016b) Using depth cameras to extract structural parameters to assess the growth state and yield of cauliflower crops. Comput Electron Agric 122:67–73. https://doi.org/10.1016/j.compag.2016.01.018

Andújar D, Rueda-Ayala V, Moreno H, Rosell-Polo JR, Escolà A, Valero C, Gerhards R, Fernández-Quintanilla C, Dorado J, Griepentrog HW (2013) Discriminating crop, weeds and soil surface with a terrestrial LIDAR sensor. Sensors (Switzerland) 13:14662–14675. https://doi.org/10.3390/s131114662

Arnó J, Escolà A, Vallès JM, Llorens J, Sanz R, Masip J, Palacín J, Rosell-Polo JR (2013) Leaf area index estimation in vineyards using a ground-based LiDAR scanner. Precision Agric 14:290–306. https://doi.org/10.1007/s11119-012-9295-0

Assirelli A, Liberati P, Santangelo E, Del Giudice A, Civitarese V, Pari L (2015) Evaluation of sensors for poplar cutting detection to be used in intra-row weed control machine. Comput Electron Agric 115:161–170. https://doi.org/10.1016/j.compag.2015.06.001

Azzari G, Goulden ML, Rusu RB (2013) Rapid characterization of vegetation structure with a microsoft kinect sensor. Sensors 13:2384–2398. https://doi.org/10.3390/s130202384

Balsari P, Doruchowski G, Marucco P, Tamagnone M, van de Zande JC, Wenneker M (2008) A system for adjusting the spray application to the target characteristics. Agric Eng Int: CIGR Ejournal X:1–11

Behroozpour B, Sandborn PAM, Wu MC, Boser BE (2017) Lidar system architectures and circuits. IEEE Commun Mag 55:135–142. https://doi.org/10.1109/MCOM.2017.1700030

Beraldin JA, Blais F, Lohr U (2010) Laser scanning technology. In: Vosselman G, Maas H-G (eds) Airborne and terrestrial laser scanning. Whittles Publishing, p 336

Berge TW, Goldberg S, Kaspersen K, Netland J (2012) Towards machine vision based site-specific weed management in cereals. Comput Electron Agric 81:79–86. https://doi.org/10.1016/j.com pag.2011.11.004

Bietresato M, Carabin G, Vidoni R, Gasparetto A, Mazzetto F (2016) Evaluation of a LiDAR-based 3D-stereoscopic vision system for crop-monitoring applications. Comput Electron Agric 124:1–13. https://doi.org/10.1016/j.compag.2016.03.017

Blasco J, Aleixos N, Roger JM, Rabatel G, Molto E, Technologies E (2002) Robotic weed control using machine vision. Biosyst Eng 83:149–157. https://doi.org/10.1016/S1537-5110(02)00155-1

Bossu J, Gée C, Jones G, Truchetet F (2009) Wavelet transform to discriminate between crop and weed in perspective agronomic images. Comput Electron Agric 65:133–143. https://doi.org/10.1016/j.compag.2008.08.004

Cheein FAA (2016) Intelligent sampling technique for path tracking controllers. IEEE Abstract 24:747–755. https://doi.org/10.1109/TCST.2015.2450180

Cheein FAA, Blazic S, Torres-Torriti M (2015) Computational approaches for improving the performance of path tracking controllers for mobile robots. Institute of Electrical and Electronics Engineers Inc., pp 6495–6500. https://doi.org/10.1109/iros.2015.7354305

Cheein FAA, Carelli R (2013) Agricultural robotics: unmanned robotic service units in agricultural tasks. IEEE Ind Electron Mag 7:48–58. https://doi.org/10.1109/MIE.2013.2252957

Cheein FAA, Scaglia G (2014) Trajectory tracking controller design for unmanned vehicles: a new methodology. J Field Robot 31:861–887. https://doi.org/10.1002/rob

Cheein FAA, Steiner G, Perez Paina G, Carelli R (2011) Optimized EIF-SLAM algorithm for precision agriculture mapping based on stems detection, vol 78, pp 195–207. https://doi.org/10.1016/j.compag.2011.07.007

Chéné Y, Rousseau D, Lucidarme P, Bertheloot J, Caffier V, Morel P, Belin É, Chapeau-Blondeau F (2012) On the use of depth camera for 3D phenotyping of entire plants. Comput Electron Agric 82:122–127. https://doi.org/10.1016/j.compag.2011.12.007

Cho SI, Lee DS, Jeong JY (2002) AE—automation and emerging technologies: weed–plant discrimination by machine vision and artificial neural network. Biosyst Eng 83:275–280. https://doi.org/10.1006/bioe.2002.0117

del-Moral-Martínez I, Rosell-Polo J, Company J, Sanz R, Escolà A, Masip J, Martínez-Casasnovas J, Arnó J (2016) Mapping vineyard leaf area using mobile terrestrial laser scanners: should rows be scanned on-the-go or discontinuously sampled? Sensors doi:10.3390/s16010119

Ehlert D, Horn HJ, Adamek R (2008) Measuring crop biomass density by laser triangulation. Comput Electron Agric 61:117–125

Erz M, Jähne B (2009) Radiometric and spectrometric calibrations, and distance noise measurement of ToF cameras. Springer, Berlin, Heidelberg, pp 28–41. https://doi.org/10.1007/978-3-642-037 78-8_3

Escolà A, Martínez-Casasnovas JA, Rufat J, Arnó J, Arbonés A, Sebé F, Pascual M, Gregorio E, Rosell-Polo JR (2016) Mobile terrestrial laser scanner applications in precision fruticulture/horticulture and tools to extract information from canopy point clouds. Precision Agric in press 18:1–22. https://doi.org/10.1007/s11119-016-9474-5

Escolà A, Planas S, Rosell JR, Pomar J, Camp F, Solanelles F, Gràcia F, Llorens J, Gil E (2011) Performance of an ultrasonic ranging sensor in apple tree canopies. Sensors 11:2459–2477. https://doi.org/10.3390/s110302459

Escolà A, Rosell-Polo JR, Planas S, Gil E, Pomar J, Camp F, Llorens J, Solanelles F (2013) Variable rate sprayer. Part 1 – orchard prototype: design, implementation and validation. Comput Electron Agric 95:122–135. https://doi.org/10.1016/j.compag.2013.02.004

Gai J, Tang L, Steward B (2015) Plant recognition through the fusion of 2D and 3D images for robotic weeding. In: American Society of Agricultural and Biological Engineers Annual International Meeting 2015

Gebbers R, Ehlert D, Adamek R (2011) Rapid mapping of the leaf area index in agricultural crops. Agron J. https://doi.org/10.2134/agronj2011.0201

Giles DK, Delwiche MJ, Dodd RB (1988) Electronic measurement of tree canopy volume. Trans ASABE 31:264–272

Gutierrez-Osuna R, Janet JA, Luo RC (1998) Modeling of ultrasonic range sensors for localization of autonomous mobile robots. IEEE Trans Industr Electron 45:654–662. https://doi.org/10.1109/41.704895

Hansen KD, Garcia-Ruiz F, Kazmi W, Bisgaard M, La Cour-Harbo A, Rasmussen J, Andersen HJ (2013) An autonomous robotic system for mapping weeds in fields. In: IFAC Proceedings Volumes (IFAC-PapersOnline). https://doi.org/10.3182/20130626-3-au-2035.00055

Llorens J, Gil E, Llop J, Escolà A (2011) Ultrasonic and LIDAR sensors for electronic canopy characterization in vineyards: advances to improve pesticide application methods. Sensors 11:2177–2194. https://doi.org/10.3390/s110202177

Maclean GA, Krabill WB (1986) Gross-merchantable timber volume estimation using an airborne lidar system. Can J Remote Sens 12:7–18. https://doi.org/10.1080/07038992.1986.10855092

Maertens K, Reyns P, De Clippel J, De Baerdemaeker J (2003) First experiments on ultrasonic crop density measurement. In: First International ISMA workshop on noise and vibration in agricultural and biological engineering 266:655–665. https://doi.org/10.1016/S0022-460X(03)00591-1

Marinello F, Pezzuolo A, Gasparini F, Arvidsson J, Sartori L (2015) Application of the Kinect sensor for dynamic soil surface characterization. Precision Agric 16:601–612. https://doi.org/10.1007/s11119-015-9398-5

Méndez V, Rosell-Polo JR, Pascual M, Escolà A (2016) Multi-tree woody structure reconstruction from mobile terrestrial laser scanner point clouds based on a dual neighbourhood connectivity graph algorithm. Biosyst Eng 148:34–47. https://doi.org/10.1016/j.biosystemseng.2016.04.013

Mukupa W, Roberts GW, Hancock CM, Al-Manasir K (2016) A review of the use of terrestrial laser scanning application for change detection and deformation monitoring of structures. Surv Rev:1–18. doi:10.1080/00396265.2015.1133039

Mutz F, Veronese LP, Oliveira-Santos T, de Aguiar E, Cheein FAA, Ferreira De Souza A (2016) Large-scale mapping in complex field scenarios using an autonomous car. Expert Syst Appl 46:439–462. https://doi.org/10.1016/j.eswa.2015.10.045

Ortega JG, Camacho EF (1996) Mobile robot navigation in a partially structured static environment, using neural predictive control. Control Eng Pract 4:1669–1679. https://doi.org/10.1016/S0967-0661(96)00184-0

Palacín J, Pallejà T, Tresanchez M, Sanz R, Llorens J, Ribes-Dasi M, Masip J, Arnó J, Escolà A, Rosell JR (2007) Real-time tree-foliage surface estimation using a ground laser scanner. IEEE Trans Instrum Meas 56:1377–1383. https://doi.org/10.1109/TIM.2007.904122

Pfeiffer SA, Guevara J, Cheein FA, Sanz R (2018) Mechatronic terrestrial LiDAR for canopy porosity and crown surface estimation. Comput Electron Agric. https://doi.org/10.1016/j.compag.2018.01.022

Piatti D, Rinaudo F, Piatti D, Rinaudo F (2012) SR-4000 and CamCube3.0 Time of Flight (ToF) cameras: tests and comparison. Remote Sensing 4:1069–1089. https://doi.org/10.3390/rs4041069

Reusch S (2009) Use of ultrasonic transducers for on-line biomass estimation in winter wheat. In: van Henten EJ, Goense D, Lokhorst C (eds) Precision agriculture '09. Wageningen Academic Publishers, Wageningen, Nederland, pp 169–175. https://doi.org/10.3920/978-90-8686-664-9

Rosell-Polo JR, Cheein FAA, Gregorio E, Andújar D, Puigdomènech L, Masip J, Escolà A (2015) Advances in structured light sensors applications in precision agriculture and livestock farming. Adv Agron 133:71–112. https://doi.org/10.1016/bs.agron.2015.05.002

Rosell JR, Llorens J, Sanz R, Arnó J, Ribes-Dasi M, Masip J, Escolà A, Camp F, Solanelles F, Gràcia F, Gil E, Val L, Planas S, Palacín J (2009) Obtaining the three-dimensional structure of tree orchards from remote 2D terrestrial LIDAR scanning. Agric For Meteorol 149:1505–1515. https://doi.org/10.1016/j.agrformet.2009.04.008

Rosell JR, Sanz R (2012) A review of methods and applications of the geometric characterization of tree crops in agricultural activities. Comput Electron Agric 81:124–141. https://doi.org/10.1016/j.compag.2011.09.007

Ruckelshausen A, Biber P, Dorna M, Gremmes H, Klose R, Linz A, Rahe R, Resch R, Thiel M, Trautz D, Weiss U (2009) BoniRob: An autonomous field robot platform for individual plant phenotyping, in: Precision Agriculture 2009. Papers Presented at the 7th European Conference on Precision Agriculture, ECPA 2009. Wageningen

Rusinkiewicz S, Levoy M (2001) Efficient variants of the ICP algorithm. In: Proceedings of international conference on 3-D digital imaging and modeling, 3DIM. https://doi.org/10.1109/im.2001.924423

Saeys W, Lenaerts B, Craessaerts G, De Baerdemaeker J (2009) Estimation of the crop density of small grains using LiDAR sensors. Biosyst Eng 102:22–30. https://doi.org/10.1016/j.biosystemseng.2008.10.003

Sanz-Cortiella R, Llorens-Calveras J, Escolà A, Arnó-Satorra J, Ribes-Dasi M, Masip-Vilalta J, Camp F, Gràcia-Aguilá F, Solanelles-Batlle F, Planas-DeMartí S, Pallejà-Cabré T, Palacin-Roca J, Gregorio-Lopez E, Del-Moral-Martínez I, Rosell-Polo JR (2011) Innovative LIDAR 3D dynamic measurement system to estimate fruit-tree leaf area. Sensors 11:5769–5791. https://doi.org/10.3390/s110605769

Schumann AW, Zaman QU (2005) Software development for real-time ultrasonic mapping of tree canopy size. Comput Electron Agric 47:25–40. https://doi.org/10.1016/j.compag.2004.10.002

Scotford IM, Miller PCH (2004) Estimating tiller density and leaf area index of winter wheat using spectral reflectance and ultrasonic sensing techniques. Biosyst Eng 89:395–408

Siegwart R, Nourbakhsh IR, Scaramuzza D (2011) Introduction to autonomous mobile robots. MIT Press, New York

Slaughter DC, Giles DK, Downey D (2008) Autonomous robotic weed control systems: A review. Comput Electron Agric 61:63–78. https://doi.org/10.1016/j.compag.2007.05.008

Tamás J, Lehoczky É, Fehér J, Fórián T, Nagy A, Bozsik É, Gálya B, Riczu P (2014) Airborne hyperspectral and LiDAR data integration for weed detection, in: EGU General Assembly 2014. Vienna

Tang J-L, Chen X-Q, Miao R-H, Wang D (2016) Weed detection using image processing under different illumination for site-specific areas spraying. Comput Electron Agric 122:103–111. https://doi.org/10.1016/j.compag.2015.12.016

Tillett ND, Hague T, Miles SJ (2002) Inter-row vision guidance for mechanical weed control in sugar beet. Comput Electron Agric 33:163–177. https://doi.org/10.1016/S0168-1699(02)00005-4

Tumbo SD, Salyani M, Whitney JD, Wheaton TA, Miller WM (2002) Investigation of laser and ultrasonic ranging sensors for measurements of citrus canopy volume. Appl Eng Agric 18:367–372

Underwood JP, Hung C, Whelan B, Sukkarieh S (2016) Mapping almond orchard canopy volume, flowers, fruit and yield using lidar and vision sensors. Comput Electron Agric 130:83–96. https://doi.org/10.1016/j.compag.2016.09.014

Walklate PJ (1989) A Laser scanning instrument for measuring crop geometry. Agric For Meteorol 46:275–284

Walklate PJ, Richardson GM, Baker DE, Richards PA, Cross JV (1997) Short-range lidar measurement of top fruit tree canopies for pesticide applications research in the United Kingdom. In: Narayanan RM, Kalshoven JE (eds) Advances in laser remote sensing for terrestrial and oceanographic applications. SPIE, pp 143–151. https://doi.org/10.1117/12.277609

Wang Q, Zhang Q (2013) Three-dimensional reconstruction of a dormant tree using RGB-D cameras. In: 2013 ASABE Annual International Meeting. ASABE, Kansas City, USA, p. Paper number 131593521. https://doi.org/10.13031/aim.20131593521

Wangler RJ, McConnell III RE, Fowler KL (1993) Application of smart submunition technology to agribusiness. In: DeShazer JA, Meyer GE (eds) Optics in agriculture and forestry. society of photo-optical instrumentation engineers - SPIE, pp 261–272. https://doi.org/10.1117/12.144035

Wei J, Salyani M (2005) Development of a laser scanner for measuring tree canopy characteristics: phase 2. Foliage density measurement. Trans ASABE 48:1595–1601

Wei J, Salyani M (2004) Development of a laser scanner for measuring tree canopy characteristics: phase 1. Prototype development. Trans ASABE 47:2101–2107

Yamamoto S, Hayashi S, Saito S, Ochiai Y (2012) Measurement of growth information of a strawberry plant using a natural interaction device. In: American Society of Agricultural and Biological Engineers Annual International Meeting 2012:5547–5556

Yan WY, Shaker A, El-Ashmawy N (2015) Urban land cover classification using airborne LiDAR data: A review. Remote Sens Environ 158:295–310. https://doi.org/10.1016/J.RSE.2014.11.001

Yandun FJ, Gregorio E, Zúñiga M, Escolá A, Rosell-Polo JR, Cheein FAA (2016) Classifying agricultural terrain for machinery traversability purposes. IFAC-PapersOnLine 49:457–462. https://doi.org/10.1016/j.ifacol.2016.10.083

Zhai Z, Zhu Z, Du Y, Song Z, Mao E (2016) Multi-crop-row detection algorithm based on binocular vision. Biosyst Eng 150:89–103. https://doi.org/10.1016/j.biosystemseng.2016.07.009

Chapter 4
Agricultural Robots for Precision Agricultural Tasks in Tree Fruit Orchards

Manoj Karkee, Qin Zhang, and Abhisesh Silwal

4.1 Introduction

Commonly accepted population growth models predict there will be more than nine billion people by 2050. Increasing population as well as an increasingly affluent lifestyle around the world have greatly increased the demand for food, water and energy. Producing enough to meet such a demand is highly challenging as farming resources such as arable land and water are limited and shrinking with increasing industrialization, urbanization and soil degradation (Karkee et al. 2009). Therefore, "improved and sustainable agricultural productivity becomes vital to meet the competing demands for food, water and energy" (Karkee et al. 2009). Production of perennial crops such as fruits and nuts has generally been found to be more sustainable in the long run compared to annual crops such as corn and wheat (Glover et al. 2010).

Amongst other resources, however, production of perennial crops such as tree fruits requires considerable seasonal labour, which is shrinking across the world. For example, the U.S. tree fruit industry depends on a large number of seasonal migrant workers (mostly undocumented), and the decreasing availability and increasing cost of labour has been a critical challenge to the long-term sustainability of this industry (Gonzalez-Barrera 2015). Mechanization, automation or robotic technologies, which incorporates sensor data collection, decision making, and precise and automated field operations, has the potential to help farmers reduce labour use, and increase their productivity and profitability. Agricultural technologies have advanced markedly over the past half-century and has led to highly automated and precise field operations, which has increased the efficiency and reduced the use of various farming inputs including chemicals, water and labour. This transformation in agricultural

M. Karkee (✉) · Q. Zhang · A. Silwal
Washington State University, 24106 N Bunn Road, Prosser, WA 99350, USA
e-mail: manoj.karkee@wsu.edu

© Springer Nature Switzerland AG 2021
A. Bechar (ed.), *Innovation in Agricultural Robotics for Precision Agriculture*,
Progress in Precision Agriculture,
https://doi.org/10.1007/978-3-030-77036-5_4

mechanization and automation, which helped reduce the farming population in the USA from 41% to less than 2% in the 20th Century (Davidson 2016), was recognized by National Academy of Engineering as one of the greatest engineering achievements (Karkee and Zhang 2012). This advancement, however, has primarily benefitted row crops such as maize and rice. For specialty crops such as fruits and vegetables, most of the field operations including harvesting, tree training and pruning, and weeding are still manual and highly labour intensive. However, there are widespread research and development efforts currently going on around the world to mechanize and auto- mate various production operations such as harvesting and selective pruning of tree fruit crops.

The adoption of automation or robotic technologies in orchards might bring social, demographic and economic changes to agricultural societies. Researchers often describe importing large numbers of seasonal migrant workers analogous to importing poverty (Gallardo and Brady 2015). The use of such technologies will undoubtedly reduce such dependency on seasonal labour in tree fruit orchards. In addition, it will create a new non-seasonal opportunity for employment to support automated or robotic farming as the maintenance and services of automated or robotic machinery will require a more skilled labour force that could be employed throughout the year. Such a change in workforce in farming operations could poten- tially minimize the seasonal fluctuation of population from a large number of low- paid migrant labourers to a smaller and more stable population of a well-paid professional workforce.

This chapter will first provide the basic understanding of crop systems in rela- tion to mechanized, automated or precision agricultural operations in orchards. The chapter will focus primarily on unique field operations in producing tree fruit crops including apples, cherries and stone fruits. Precision, automation and mechanization technologies being used in field operations such as planting and chemical application are introduced briefly (Sect. 4.2), whereas the technologies to be discussed in detail include tree pruning, flower and green fruit thinning, and harvesting (Sects. 4.3–4.5). Finally, the potential direction for future research and development will be discussed that can lead to the next level of automation and robotic solutions for tree fruit orchards.

4.1.1 Tree Fruit Planting Systems

Tree fruit crops such as apples, cherries, pears and stone fruits are planted and trained in various ways that create different types of canopy architectures. Canopy architecture, which is the organization of trunks, branches and other components in a three-dimensional (3-D) space is altered over time and space by training and pruning operations. Traditionally, comparatively bigger trees have been grown without any trellis support (also called conventional canopies). These free standing trees could be planted several metres apart within and between rows (a few hundred trees per hectare) and can grow several metres tall and wide. Conventional canopies are also

Fig. 4.1 A conventional
cherry tree architecture
(from Amatya 2015)

called four-dimensional canopies with three geometric dimensions and variation between individual trees as the fourth dimension. Branches are often allowed to grow in any direction, but are pruned to create a pyramidal structure allowing good light distribution to all levels of the canopy (Fig. 4.1). These conventional canopies present challenges both for manual operations by requiring use of tall ladders placed at various locations around the canopy and for robotic operations by limiting the visibility and accessibility to various parts of the canopy.

To increase worker productivity and simplify the work environment for automated or robotic machines, horticulturists and growers have been developing and planting more Simple, Narrow, Accessible and Productive (SNAP) orchards in recent years. These modern orchards, which are increasingly replacing conventional orchards, use trellis wire to support smaller, dwarf trees that are planted much closer together leading to several thousand trees per hectare in general (e.g. Lang et al. 2015). Branches are trained and pruned such that canopy depth remains very narrow, thus creating a fruiting wall structure (Fig. 4.2).

Several variations of SNAP tree architectures have been developed and planted for different types of crops. Some canopy architectures are a logical progression of central leader conventional trees such as tall spindle apples (Fig. 4.3) and super slender cherries (more in Karkee et al. 2018). Fruiting wall canopy architectures have

Fig. 4.2 Aurora apples grown in a fruiting wall architecture in a commercial orchard, Washington State, USA

also been designed using multi-leader trees. Upright fruiting offshoot (UFO) cherries (Whiting 2008) and bi-ax apple trees are some of the examples of multi-leader fruiting wall systems.

4.2 Introduction to Orchard Operations

Production of tree fruit crops such as apples, cherries, pears and stone fruit involves several field operations including *crop establishment* (e.g. site selection, land preparation and planting), *input application* (e.g. irrigation, cooling, nutrients and pesticides), *crop monitoring, canopy and crop-load management* (e.g. pruning and training, pollination and thinning), *pest control* (e.g. weed, bird and other animal controls), and *harvesting and field logistics.* In the following sub-sections, various pre-harvest orchard operations are discussed briefly. In addition, tree training and pruning, thinning, and harvesting and logistic technologies for tree fruit crops are discussed in Sects. 4.3–4.5.

Fig. 4.3 A tall spindle apple orchard in Washington State with around 7,000 trees per hectare

4.2.1 Crop Establishment

As listed above, some of the major tasks in tree fruit orchards for crop establishment include site selection, land preparation and planting. *Site selection* is primarily based on availability of water, soil type, weather and topography (e.g. altitude, slope and slope aspects) (Seo 2003). Geographic Information System-based modelling is widely used in selecting desirable sites for developing new orchards or sometimes to change the type of crops based on the market trend or suitability of the site (Thomas et al. 2002). Precise *planting* of trees in a desired direction, row spacing and tree spacing is achieved using automated planting machines steered using Real Time Kinematic (RTK) GPS technology with centimetre level positioning accuracy. Some machines offer fully automated (or robotic) planting in which digging a hole, separating and dispensing a plant from a bundle, putting it in the hole and filling the hole with soil is all completed by the machine. With other machines, trees are planted manually into holes prepared by automatically steered machines.

4.2.2   Crop Input Application

Some of the important inputs for tree fruit crops include water, fertilizer and other chemicals used for water and nutrient management, disease and insect control, plant growth regulation, and flower and green fruit thinning. *Precision irrigation* is crucial to provide the right amount of water at the right time and right location both to improve crop yield and quality while minimizing total water use. Sensor-based automated irrigation scheduling has been and will become an even more important component of an overall orchard automation system because water is becoming one of the most limited resources for farming. Considerable effort has gone into using weather data, evapotranspiration information and soil and or plant sensing techniques to estimate variable water demand in different parts of an orchard and to apply water precisely to meet the variable demand (Fernández and Cuevas 2010). Researchers have also been developing deficit irrigation strategies as well as sub-surface drip irrigation systems that could bring water directly to root zones, thus reducing evapotranspiration and other losses and increasing water use efficiency.

In various growing regions, fruit *cooling* can be important to minimize sunburn to fruit. Evaporative cooling through overhead water application is one of the common techniques used (Parchomchuk and Meheriuk 1996). In recent years, netting or shade cloths or fabric have also been used to minimize sunburn while also providing a pest control solution.

Chemical application technology is used widely for all types of crops to apply nutrition, pesticides, plant growth regulators and other chemicals to keep plants healthy and productive. In orchards, different types of air-blast sprayers are commonly used for chemical application. Chemigation or Fertigation (application of nutrition or other chemicals through an irrigation system) and 'solid set' canopy delivery (installation of a fixed hydraulic system in the orchards) are some other methods available or being investigated for the application of chemicals in orchards (Sharda et al. 2015). Different levels of automation and precision are possible with different types of machines. Details on these technologies can be found in Karkee et al. (2013).

4.2.3   Canopy and Crop-Load Management

Field operations for canopy and crop-load management include pruning and training, pollination and thinning. Tree *training and pruning* are two important cultural practices performed to create the desired shape and size of tree canopies to optimize light distribution to different parts of the canopy, to simplify the canopy for both manual and machine operations, and to keep the trees healthy and productive (Karkee et al. 2014). In modern orchards, trees are trained to trellis wires starting the same or following year of planting to create very narrow two dimensional structures also called fruiting walls (Sect. 4.1.1).

Every year, a certain proportion of branches are pruned out during the dormant period or growing season, and new branches are trained to the desired location and orientation. This operation renews the canopy by removing older, diseased and otherwise unproductive branches while keeping the newer and healthier branches for fruit production. Training and pruning that are done well also help flower and fruit to grow in such a way that robotic thinning and harvesting could be more effective.

Crop *pollination*, the process of transferring pollen to stigma of the flower from the male anther in modern orchards is a complex phenomenon that involves well-planned coordination between pollinizers, pollinators (pollinating animals or tools) and pollens. Pollinizers are generally the trees planted in the orchard that are of the same species but of a different variety from the crop trees. Pollinators are generally the biological agents such as honey bees that transfer pollen to the stigma through their movement between pollinizer and regular crop trees.

Pollination can also be performed mechanically using a carefully designed pollen medium with a certain level of pollen concentration, which is applied to tree canopies using a traditional sprayer or a robotic machine like a drone (Sutyemez 2011). Mechanical pollination, although still in its early stages of research and development, has the potential to revolutionize the fruit crop pollination system by avoiding the need for the decreasing population of pollinators (e.g. honey bees). The method also provides an opportunity for precise (temporal and spatial) pollination, which can ensure fertilization occurs at the right time and right amount. If precision pollination can be achieved, it could reduce the need for bloom or green fruit thinning while also improving fruit quality and yield.

In general, more fruit is set than a tree can reasonably bear to grow the desired quality fruit. To minimize the possibility of setting a large amount of fruit, a proportion of the flowers are thinned mechanically or chemically in various crops including cherries and stone fruit. In some intensively managed orchards with 'high end' produce, flower thinning is completed manually. Because there is a risk of not having enough fruit set, bloom or flower thinning is generally performed conservatively, which often leads to the need for green fruit thinning. Green fruit thinning can also be done by chemical and mechanical methods, but finishing with hand thinning is desirable to make sure the right amount and distribution of fruit are achieved. Thinning is also essential to facilitate mechanized or robotic harvesting of crops like apples and peaches by avoiding clusters of fruit and preventing fruit from growing in difficult positions such as behind a tree trunk.

4.3 Robotics for Tree Pruning

Because of their wide biological diversity, there are many ways to train and prune fruit trees. In the Pacific Northwest (PNW) region of the USA, training is often performed by manually tying tree branches to certain trellis wires or other structures. Tree pruning is also completed manually by physically cutting and removing parts of a tree following some subjective and quasi-objective guidelines. Hand training and pruning are highly labour intensive operations, and also require workers to have adequate knowledge and skill to do the job well. Use of some kind of powered tools, such as pneumatic, hydraulic or electric scissors, could reduce the force required to cut large sized branches, which is important to reduce worker fatigue and improve their productivity during long working hours. To reach higher canopy regions, workers commonly use ladders. Orchard platforms have also been used to assist workers to reach desired work areas in the canopies safely and with minimum effort.

Modern fruiting wall architectures (Sect. 4.1.1) are characterized by high tree densities requiring intensive training and pruning. Therefore automated pruning is crucial to achieve the desired level of precision and to keep labour use and associated costs to a sustainable level. In an automated or robotic pruning operation, unproductive and diseased branches and shoots will be removed selectively from trees. However, selectively pruning a fruit tree requires a high level of intelligence, which makes it a very challenging task to meet pruning quality requirements satisfactorily with a machine. To solve this problem, researchers have worked to create objective rules for pruning decisions based often on subjective knowledge and skill that experienced human workers and horticulturalists use in fruit tree pruning.

In addition, modern canopies are trained to minimize or avoid the subjectivity of deciding which branches to prune out every year. Such simplification not only helps manual labour to be more productive and precise in pruning fruit trees, but also opens up the opportunity for automated or robotic pruning using a few objective rules. For example, UFO (Upright Fruiting Offshoots) sweet cherry architecture was designed such that only vertical limbs of a horizontally trained trunk are allowed to grow. Any secondary branches grown out of those vertical offshoots are pruned out. With this type of simplification, pruning crews can be trained to perform the job very accurately while also providing simplified decision making rules for a robotic harvester. In apple orchards, formal fruiting wall and tall spindle orchards enable a simplified set of rules to be provided for pruning trees. For example, more than 95% of pruning in a tall spindle orchard can be accomplished using the following two rules (Karkee et al. 2014).

1. Prune out long or large branches (longer or larger than a threshold size that the user or grower can define)
2. Prune out one of the two closely spaced branches (threshold spacing can be provided by growers)

A robotic pruning system will have to incorporate a sensing or machine vision system to generate the 3-D structure of trees, and include capabilities for identifying branches to be pruned on fruit trees using pruning rules or strategies. These capabilities will then be integrated with a robotic manipulator and end-effector system to complete a selective pruning operation. Researchers have long been studying robotic components and systems for selective pruning. Sevilla (1985) used modelling and simulation to study a robotic manipulator for grapevine pruning. The manipulator was tested in the laboratory environment, and it was concluded that robotic pruning of grapevines was technically feasible. Ochs and Gunkel (1993) also worked on a machine vision system for a grapevine pruner. Similarly, Lee et al. (1994) reported

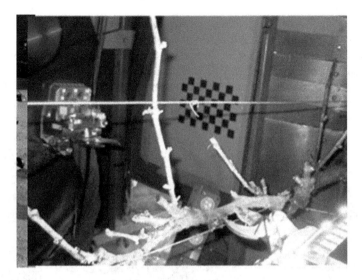

Fig. 4.4 A robotic grape vine pruner demonstrated by Vision Robotics (from visionrobotics.com)

research on the electro-hydraulic control of a vine pruning robot. Kondo et al. (1994) developed a manipulator and vision system for a multi-purpose vineyard robot. Some private companies have also investigated robotic pruning solutions with a primary focus in pruning grapevines. For example, Vision Robotics (http://www.visionrob otics.com/) has worked on grape vine pruning for several years (Fig. 4.4) and has demonstrated the technology in recent years.

4.3.1   Sensing Systems for Robotic Pruning

A proper sensing technique is essential to create the 3-D structure of fruit trees and identify unwanted branches for robotic pruning. Machine vision technology can be used to obtain such information non-destructively. In the past, some researchers investigated the use of machine vision to locate and follow grapevines using digital image processing to guide an automated pruning device (Naugle et al. 1989). The image processing to identify cordons was performed as a binary image obtained after the thresholding of colour images recorded by a single camera. Gao and Lu (2006) also used a single camera for identifying pruning points in grapevines. Extracting 3-D information on objects from a single image is difficult, but such representation of objects and scene is essential in developing such automatic machines. Researchers have been attracted by 3-D technologies such as stereo-vision and a Time-of-flight of Light (ToF)-based 3-D camera (called *3-D camera* in the following text) for 3-D reconstruction. Three-dimensional (3-D) reconstruction of the object of interest is essential in such applications. Beder et al. (2007) found that a time-of-flight-of-light-based 3-D camera could provide better reconstruction accuracy than a stereo-vision system, but a stereo-vision system has the potential to provide higher resolution data. Laser sensors have also been widely used in acquiring 3-D information.

The WSU Center for Precision and Automated Agricultural Systems has also been conducting research on the 3-D reconstruction of apple trees and defining and using objective rules for pruning point identification with a robot (Fig. 4.5). They used 3-D

Fig. 4.5 a A 3-D sensing system in a commercial apple orchard and **b** the outcome of a pruning branch identification method investigated by researchers at Washington State University (Karkee et al. 2014; Karkee and Adhikari 2015)

(a)

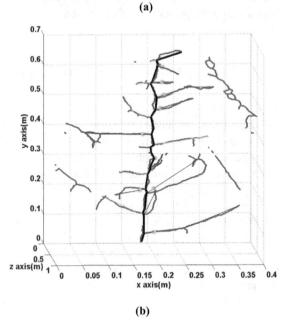

(b)

camera-based point cloud acquisition and a skeletonization method to create the 3-D structure of tall spindle apple trees with a central trunk and lateral branches (Karkee and Adhikari 2015). Human experts were interviewed to understand how they make decisions for apple tree pruning. In addition, the actions of experienced pruners were recorded and analysed. It was found that, for the tall spindle tree architecture studied, the pruning process can be captured by four basic rules (Karkee et al. 2014):

(1) *Remove diseased branches,*
(2) *Remove long branches*
(3) *Remove large branches*
(4) *Remove closely spaced branches.*

Pruning rules were then used to identify branches to be pruned out. Field experiments with automated pruning point identification showed comparable results with a group of experienced pruners (Karkee et al. 2014). When only two pruning rules (second and fourth) out of the four described above were used, the algorithm suggested a branch removal of 19.5% of total branches, whereas in the same situation human workers suggested 22% removal (Karkee et al. 2014). This study also revealed various challenges including: (1) detection of all required targets, such as dead branches, with machine vision was difficult and (2) identifying pruning points and steps in complicated canopy structures often required judgment. In such complex situations, robust results could potentially be achieved through human–robot collaboration, where the human performs tasks that require judgment and robots perform tasks with rule-based decisions.

A group of researchers led by Peter Hirst at Purdue University investigated various ways of 3-D reconstruction of grape vines and fruit trees. They proposed a new 3-D reconstruction method for apple trees trained in a tall spindle architecture (Elfiky et al. 2015). They used a Kinect 2 sensor (a gaming sensor from Microsoft, quite inexpensive), which showed potential for a low cost sensor for orchard machine-vision applications. These efforts at various universities and private companies have laid good foundations for further research and development in automated pruning of tree fruit crops.

4.3.2 *Potential for Automated Training*

Branch detection methods developed for automated pruning or harvesting might also provide a starting point for automated training. Some simpler canopy structures provide potential for automated training using such branch detection techniques. For example, a UFO cherry orchard can be trained by tying vertical offshoots to trellis wires as they grow past individual wires. However, in general, training is a more challenging problem in terms of the complexity of decision making and handling of generally young branches. Because of these challenges, research and development in mechanization or automation of fruit tree training has been limited.

4.4 Automation and Robotics for Thinning in Orchards

Blossom and green fruit thinning are common practices of crop-load management in tree fruit production for yielding larger and or otherwise better quality fruit. Thinning can be achieved by hand or chemical thinning at bloom or post bloom (green fruit stage), or mechanical crop thinning, which is commonly used at both blossom and green fruit stages. Hand thinning at either stage could reliably achieve the thinning goal, but is a costly and labour intensive operation. Chemical thinning is an alternative to hand thinning, however, the efficacy of chemical thinning depends on the weather conditions, fruit variety, tree age, time of application and chemical concentration, as well as being subject to increasing pressure for environmental protection and more strict regulations for chemical registration. Mechanical thinning has been explored well in the past as a potential alternative and has shown promise as an effective option for tree fruit thinning.

One issue both chemical and mechanical thinning methods face is the considerable variation and uncertainty in the results caused by various known and unknown factors such as environmental and weather conditions, cultivar, canopy density, canopy type and instantaneous density of blossom or green fruit. This issue could potentially be addressed using a robotic thinning system. Machine vision and other sensing systems can be used to detect both flowers and green fruit in orchard environments and locate target flowers or fruits for thinning (e.g. Lyons and Heinemann 2016). A robotic end-effector such as a brushing head or chemical injector could then be manipulated to reach the target and destroy the desired amount of flowers or green fruits.

The overall development of automated or robotic thinning is still in its infancy. The possibility of an automated or robotic machine for blossom thinning is an area of recent research and development, with most currently reported efforts focusing on developing sensing, machine vision, manipulator and end-effector technologies for robotic thinning operations (Emery et al. 2010; Lyons and Heinemann 2016). Lyons and Heinemann (2016) have invented a selective automated thinning technique, which uses a string thinner with a number of adjustable thinning heads to perform automated, targeted thinning operations. A more precise operation will be essential for thinning at the individual blossom or green fruit level or small cluster level as is achieved by hand thinning. More comprehensive research and development in all aspects of robotic thinning is essential before this technology could become commercially adoptable.

4.5 Automation and Robotics for Tree Fruit Harvesting

Fresh market tree fruit crops such as apples and cherries are harvested manually around the world. In Washington State alone, more than 16 billion apples are produced annually, which means that a human hand needs to repeat a specific harvesting motion 16 billion times over the harvesting window of about 10 weeks in the state (Fig. 4.6).

Fig. 4.6 Manual apple harvesting in a modern SNAP (simple, narrow, accessible, productive) orchard, in Prosser, Washington, USA

Sweet cherry, pear and stone fruit are also large crops in this region of the USA, together requiring more than 30 thousand seasonal workers to pick these tree fruit crops. In addition to challenges of dwindling labour availability, harvesting cost is a major component of total tree fruit production costs. For example, apple harvesting in WA orchards can cost more than $5,500 per hectare per year (Galinato and Gallardo 2011). Worker health and safety is another large concern in tree fruit harvesting. In hand-harvesting, workers use ladders to reach upper parts of the canopies, which requires repeated climbing up and down the ladder with a big load of fruit. Workers continue to pick fruit while they climb the ladder up or down, which further increases the risk of fall, injury and even death.

To minimize dependence on seasonal labour and reduce harvesting costs, universities, research institutions and private companies have been working on developing various types of mechanized and automated harvesting solutions for a long time. Principally, there are two methods that have been investigated for fresh market tree fruit harvesting; shake-and-catch (can be called mass harvesting) methods and pick-and-place (can be called robotic harvesting) methods. These methods are described below including an overview of various efforts around the world to develop individual techniques. More detail on research into tree fruit harvesting and developments around the world can be found in Karkee et al. (2018).

4.5.1 Mechanized or Mass Harvesting

Mechanized or mass harvesting of fruit crops is generally achieved by applying vibratory motion to the canopy, trunk or targeted branch(es) to detach fruit, which are then collected on various types of catching surfaces. This technique is also called shake-and-catch harvesting. There are various ways that shaking can occur including inertial shaking, repeated impacting of tree trunks or even air blasting. Inertial-type harvesters basically use an oscillating mass to accelerate tree canopy and fruit to a level when the inertial force developed surpasses the retention (or bonding) force between the fruit and the tree branch (Pacheco and Rehkugler 1980). In impact shaking, an impulse signal is applied to the trunk or limbs of the tree, which causes rapid acceleration of the fruit leading to their detachment (Pacheco and Rehkugler 1980). Researchers have also investigated the use of continuous shaking of fruit like apples and cherries (Zhou et al. 2014). Fruit response studies have shown that fruit detachment during continuous shaking occurs after the fruit has developed an oscillatory motion with an adequate frequency or amplitude to generate sufficient force for detachment (Diener et al. 1965).

Fruit catching is considered one of the most critical components of a shake-and-catch harvesting system because it determines both the percentage of fruit that can be caught as well as fruit quality that can be achieved with the harvesting system. Various types of catching surface materials and mechanisms have been studied including different types of foams and foam-based buffering mechanisms (Ortiz et al. 2011; He et al. 2013), non-Newtonian fluid (De Kleine and Karkee 2015) and air suspension systems (Ma et al. 2016).

Researchers have long been developing and evaluating shake-and-catch harvesting systems for various types of nuts and tree fruit crops such as berries, cherries and citrus fruits (Parameswarakumar and Gupta 1991; Peterson and Wolford 2003; Polat et al. 2006; Karkee et al. 2016). Further discussion on these efforts is avoided here as the chapter focuses primarily on robotic approaches for orchard crops. Interested readers can refer to Karkee et al. (2018) and other references for further details.

Shake-and-catch systems could achieve a much faster harvest speed or productivity (amount of fruit harvested per unit time) compared to human or robotic picking because a large number of fruit could be detached and collected within a few seconds using this technique compared to one fruit every 2 to 3 s with manual picking or with some of the best robotic harvesters developed so far (Davidson 2016; Sect. 4.2.2). However, commercial adoption of shake-and-catch harvesting techniques for fresh market tree fruit crops has still been lacking to date, primarily because of excessive fruit damage, and insufficient removal efficiency or robustness of the machines.

4.5.2 Robotic Harvesting

As discussed in Sect. 4.5.1, mass or bulk harvesting methods have shown some promise for harvesting fresh market tree fruits. However, the techniques still require more comprehensive research on system optimization and socioeconomic analysis before they would be ready for commercial adoption. For crops with greater sensitivity to bruising, robotic harvesting systems that can handle fruit more gently are essential. Even though the complexity of robotic systems might increase the cost of automated harvesting, higher gross revenue in many of the high-value crops (e.g. Honeycrisp apples) offers the potential for commercial adoption even with relatively larger capital investment. In this sub-section, past accomplishment and recent trends in research and development for robotic tree fruit harvesting are discussed.

Over the last few decades, researchers have studied robotic systems for harvesting different types of tree fruit crops including apples and citrus fruits (Harrell et al. 1990; Rabatel et al. 1995; Baeten et al. 2007). The MAGALI project was completed in the 1980s in France with a goal to develop a tree fruit harvesting robot. The team evaluated the prototype machine with apples achieving a harvesting efficiency of ~50% (D'Esnon et al. 1987). The robot took 4 to 9 s to harvest a fruit. Harrell et al. (1990) also developed a robotic harvesting system and evaluated the machine in citrus harvesting in France and Florida. The robotic system was able to detect about 75% of total fruit and harvest each of those fruit in about 3 to 7 s. The EUREKA project for Citrus harvesting was completed in Spain and France in the 1990s with a fruit detection accuracy of 90% (Rabatel et al. 1995).

Agribot, a project in the 1990s in Spain, focused on human and machine collaboration in completing harvesting. They used a gripper-cutter end-effector, which took about 2s to detach a fruit. Muscato et al (2005) studied an orange picking robot using several simulation tests, which showed an average picking time of 5.93 s per fruit. Baeten et al. (2007) developed an Automated Fruit Picking Machine (AFPM) and evaluated it with apples in Belgium (Fig. 4.7). The machine could harvest about 80% of the fruit with a cycle time of roughly 9 s.

More comprehensive reviews on various components of robotic fruit harvesting systems are available in Gongal et al. (2015), and Davidson (2016). As discussed earlier and summarized in Fig. 4.8 and Table 4.1, past efforts have achieved variable and inconsistent cycle times (time taken to harvest each fruit) for fruit harvesting. Even faster prototypes developed in the past fall behind the speed of manual fruit picking (approximate cycle time of 2 s).

Because of the lack of the desired level of speed, accuracy, robustness and cost, no commercial success has been achieved in robotic tree fruit harvesting. However, the problem has become more important recently as the cost and availability of labour has become increasingly challenging (Gonzalez-Barrera 2015). Therefore, there has been renewed interest by the tree fruit industry, governments, academia and private companies in developing robotic harvesting solutions for fresh market tree fruit crops. A company in New Zealand (Robotics Plus) has been working on commercially adoptable machines for kiwi and apple harvesting (Fig. 4.9, http://www.roboticsplus.co.nz/

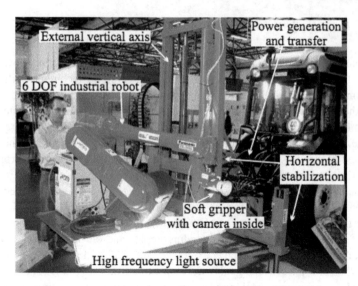

Fig. 4.7 Automated Fruit Picking Machine (AFPM) (from Baeten et al. 2007)

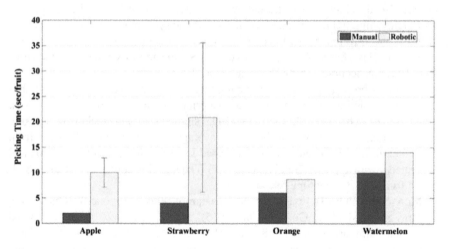

Fig. 4.8 Single fruit picking time (Cycle time) of various robotic harvesting machines compared to manual picking time for corresponding fruit crops (from Davidson 2016)

our-work). Abundant Robotics (http://www.abundantrobotics.com/) and FFRobotics (http://www.ffrobotics.com/) are some of the other companies working on developing robotic machines for commercial apple harvesting through public, private or venture capital funding.

Thomas Burks and his team at the University of Florida have also been working on robotic citrus harvesting systems with novel control algorithms for robust fruit picking. *Clever Robots for Crops* is another recently completed project funded by

Table 4.1 Agricultural Harvesting Robotics Projects: 1986–2012 (adapted from Davidson 2016). Question mark '?' in the table (under picking time) indicates that there is some uncertainty with the time reported

Fruit	Study	Manipulator's actuator	End-Effector 's actuator	Picking time (Sec)	Detachment success rate (%)	Damage rate (%)
Apple	D' Esnon et al. (1987)	Hydraulic	Vacuum	4(7)	50	25
Apple	Setiawan et al. (2004)	Electric (Industrial)	Vacuum generator	11.9	–	–
Apple	Baeten et al. (2008)	Electric (Industrial)	Vacuum pump	8(?)	80	30
Apple	Bulanon and Kataoka (2010)	Electric (Custom)	DC & stepper motors	7.1	90	–
Apple	Zhao et al. (2011)	Electric and Hydraulic (Custom)	DC motor and Pneumatic pumps	13 (?)	77	–
Cantaloupe	Edan et al. (2000)	Electric (Custom)	Pneumatic (?)	15	85	7
Cherry	Tanigaki et al. (2008)	Electric (Custom)	Servomotors and Vacuum pump	14 (?)	78	43
Grape	Monta et al. (1995)	Electric (Custom)	–	–	–	–
Kiwi	Scarfe et al. (2009)	Electric (Custom)	Electric motor	1	–	–
Orange	Pool and Harrell (1991)	Hydraulic (Custom)	Hydraulic	3–7 (?)	69	44
Orange	Muscato et al. (2005)	Hydraulic and pneumatic (Custom)	Pneumatic	8.7	–	–
Orange	Lee and Rosa (2006)	Hydraulic and pneumatic (Custom)	Pneumatic cylinder	–	84	–

(continued)

Table 4.1 (continued)

Fruit	Study	Manipulator's actuator	End-Effector 's actuator	Picking time (Sec)	Detachment success rate (%)	Damage rate (%)
Tomato	Monta et al. (1998)	Electric (Custom)	DC motor and vacuum pumps	15	91	–
Tomato	Ling et al. (2004)	Electric (Custom)	Vacuum pump, stepper motor, linear actuator	–	–	–
Strawberry	Hayashi et al. (2010)	Electric (Custom)	Pneumatic	31.3 (?)	86	–
Strawberry	Han et al. (2012)	Electric (Custom)	Electric motors	–	–	–
Sweet Pepper	Kitamura and Oka (2005)	Electric (Custom)	DC motor	–	–	–
Watermelon	Umeda et al. (1999)	Electric (Custom)	Vacuum pump	–	66	–
Watermelon	Hwang and Kim (2003)	Electric (Custom)	Pneumatic	14 (?)	–	–
Watermelon	Sakai et al. (2008)	Hydraulic and Electric (Custom)	DC motor	14	86	0

Fig. 4.9 A Kiwi fruit harvesting robot developed by Robotics Plus

the European Union to develop robotic harvesting system for apples and other crops (Eizicovits and Berman 2014). This team was later funded to develop a commercially adoptable robotic system for fruit crops (Barth et al. 2016).

Another effort to develop a robotic apple harvesting system is at Washington State University (WSU). The team has focused on: (i) novel machine vision approaches to improve the accuracy of fruit detection to almost 100% (Silwal et al. 2016), (ii) improving the picking end-effector design by understanding manual picking dynamics (i.e. the knowledge of how the human hand moves during picking and how much force is applied by each finger over time) (Davidson et al. 2016) and (iii) developing a low cost manipulator that can meet specialized specifications for apple harvesting in fruiting wall orchards (Davidson 2016). The integrated prototype robot (Fig. 4.10) developed by the team has achieved a fruit removal efficiency of 85% with a cycle time of about 6 s. The team has also been working on dual manipulator coordination for picking and collecting apples that minimizes the travel distance, which therefore has the potential to improve overall fruit picking cycle time (Silwal et al. 2016).

Given the need and support from the tree fruit industry around the world and several funding programs around the world targeted to innovation in agriculture, it is expected that these efforts in both private and public sectors will continue and lead to one or more types of mechanized, automated or robotic machines becoming commercially available for tree fruit harvesting in the near future.

Fig. 4.10 A prototype robotic apple harvesting machine developed at Washington State University is being evaluated in a commercial fruiting wall orchard (Prosser, Washington, USA)

4.5.3   Fruit Handling in Orchards

Bin management, which involves bringing in empty containers (bins) and removing filled bins from harvesting zones, is an important orchard operation during harvesting. Typically, pickers pick fruit and unload it into nearby bins at regular intervals before harvesting begins. As harvesting progresses, the harvesting zones will gradually deviate from bin locations, and when the pickers move away from them, they might need to be moved closer to the pickers. Once a bin has been filled, it is removed (using a human operated forklift type of machine) from the harvesting zones. Bin placement and removal are usually done by tractor-mounted forklifts or bin trailers in individual trips.

In a typical apple orchard of the US PNW region, it is common to have 75 to 100 metric tons per hectare of crop, which could go up to 120 or more metric tons in some high density orchards. Because of the large number of containers or bins required to move this level of production out of orchards, increasing their handling efficiency is essential for improving the overall fruit harvesting process. Hood et al. (1981) have used a tractor and bin trailer to create a fruit handling system capable of carrying four bins at a time. Hedden et al. (1984) designed a fruit handling system that used a truck-mounted basket elevator to manage the bins. These or similar systems are commercially available (e.g. www.bluelinemfg.com) and have been adopted widely by the tree fruit industry, but it is difficult to find recent literature suggesting improvement or new designs of bin handling machines.

To reduce dependence on human labour during the harvest season and further improve overall harvest efficiency, Washington State University researchers have proposed and validated the use of a robotic machine for simplifying bin handling (Ye et al. 2017). The robot was designed with a 'go-over-the-bin' feature which allows the bin handler to go over empty bins to pick up a full one rather than going around the row to reach the full bin. With this type of over-the-bin movement, the bin movement and replacement process can be completed faster and can reduce the waiting or travel time of the harvesting crew (manual or robotic in the future) (Fig. 4.11).

The machine uses a four-wheel-independent-steering system, which allows the robot to be maneuvered effectively and accurately within the confined space of a tree row to complete bin picking, placing and delivering tasks. Because the ground surface in commercial orchards is often uneven and could frequently be wet and muddy, a wheel–ground engaging system was used to ensure all wheels are reliably engaged with the ground surface. Similarly, a reliable and effective bin handling system requires accurate positioning and orientation of the robot for bin picking and placing by a navigation control system.

Fig. 4.11 An automated bin handling machine (also called automated 'Bin Dog') developed at Washington State University

4.6   Opportunities and Challenges

Energy, economic and environmental sustainability are crucial for the success of any agricultural production system in the future. Reduction of farming inputs (including labour), optimization of crop yield and quality, and protection of the environment (e.g. reduce level of chemicals introduced into the environment) will all become important. Automation and robotics have played and continue to play an important role in achieving these often competing goals in farming.

Although there has been some success in mechanization and automation, production of specialty crops such as tree fruits and vegetables still remains labour intensive. Labour availability is uncertain at best and costs have been steadily increasing. In addition, there are more stringent regulations on the quality and safety of labour that reinforce the need for more automated production. Without new mechanization and automation solutions in the next five to ten years, the tree fruit industry could suffer greatly.

The tree fruit industry, local and central governments, and public and private institutions around the world have realized that efforts need to focus on developing new mechanization and automation technologies to deal with the serious issues facing specialty crop production in general, and tree fruit production in particular. If these efforts continue, highly automated tree fruit production can become a reality. There are a few important areas of research and development that scientists can focus on to bring these technologies to a new level so that commercial adoption of mechanized, automated and robotic technologies for tree fruit production becomes a reality sooner rather than later.

Multi-purpose Machines: Considerable effort has been given to robotic and shake-and-catch harvesting, therefore, it can be anticipated that these machines will be ready for commercialization in the near future. Because automated or robotic machines will be complex and expensive, a modular design is essential so that a machine can be used for multiple field operations over the year. For example, modules could be developed for harvesting, pruning and thinning that could be installed in a single base robotic machine.

Multi-machine Collaboration: As automated machines or robots become viable for field operations, it is essential to have more than one in a field to complete time sensitive tasks. This could be for multiple robotic hands in an apple harvesting machine, a number of robotic Unmanned Aerial Vehicles (UAVs) flying to deter birds, and so on, but efficient and effective cooperation and collaboration between these machines is essential. Research and development will be important so that interference, work space overlap and collision between machines could be avoided while ensuring timely completion of field operations.

Human Machine Collaboration: As famously stated elsewhere, getting 80 to 90% of the job done with a machine could be achieved at reasonable complexity and cost, but completing 95 to 100% of a field operation by a robotic machine might require substantially greater machine complexity and capability, and therefore associated cost. Therefore, effective and efficient partnership between human and machine could be a viable path for automating many of the still-manual field operations in tree fruit production. Humans and machines can complement the capabilities of each other to achieve the desired level of labour saving while keeping the machine's complexity and cost to an acceptable level.

Trans-disciplinary Approaches: First, automated technologies for tree fruit crops are generally adapted from existing technologies such as mechanical manipulators and machine vision systems used in industrial robotics. Adoption of such technologies into agriculture often expose them to challenging work requirements and environmental constraints present in outdoor agricultural fields. Even though specialized solutions are designed for specific agricultural operations, engineering and automated solutions developed in isolation of other crucial aspects of production agriculture are generally destined for failure. For example, improvement in existing fruiting wall canopy architectures or new canopy designs that can facilitate mechanized and automated field operations (e.g. improved visibility and accessibility to fruit for robotic harvesting) is crucial for improving the robustness and reducing the cost of automated solutions. Collaboration with socioeconomic scientists is equally important so that the new technologies can be adopted with the least possible impact on the socio-cultural structure of the communities while ensuring economic viability for the growers.

4.7 Summary and Conclusion

Global population is expected to grow beyond 9 to 10 billion in the next 3 to 5 decades. In addition, 3 to 4 billion people are expected to become more affluent demanding significantly more resources to support the changing lifestyle. These factors lead to an exponential growth in demand for food, fibre and fuel. Producing enough with decreasing farming resources is, indeed, the greatest challenge of our generation, and ironically, of only our generation. If this generation is able to meet the production challenges sustainably by the next 3 to 5 decades, the next and future generations may not have to deal with this challenge to the same extent because the world population will plateau or even go down slightly by then as suggested by various population growth models.

Labour has been one of the most limiting farming resources for sustainable tree fruit production in many parts of the world. In the USA, the majority of the migrant workers (approximately 70%) come from Mexico (USDA Economic Research Service 2010). According to the Farm Labor Survey (FLS) of the National Agricultural Statistics Service (NASS), the average number of farmworkers including agricultural service workers decreased from 1.142 million to 1.063 million from the year 1990 to 2012. A similar challenge of labour availability and associated cost is faced in other fruit growing regions including Europe, Australia, New Zealand and Japan as well as Chile, China, India and Brazil. Therefore, the lack of mechanized, automated and or robotic systems threatens the long-term sustainability of fresh market tree fruit crop production in the USA and around the world.

Recognizing this challenge, research institutions and private companies around the world have long been investigating various tools, techniques and methods for mechanizing or automating various field operations in tree fruit production including land preparation and planting, chemical application, pest control, training and pruning, flower and green fruit thinning, and harvesting. These efforts have led to some successful mechanization and automation technologies that have reduced labour use, improved their health and safety, and improved overall fruit yield and quality. Some of those technologies that have been successfully applied to tree fruit production include RTK-GPS-based planting and variable-rate chemical application systems. However, the bulk of the most labour-intensive tasks such as fruit harvesting and tree pruning remain manual. In Washington state, USA, about 30,000 seasonal workers are needed during tree fruit harvesting season. It is increasingly challenging every year for farmers to hire and manage enough seasonal workers so that the valuable crops can be harvested at the right time. On the other hand, such huge seasonal employment can create huge socio-economic challenges for the towns and cities in fruit growing reasons.

In recent years, the tree fruit industry, and local and central governments in key fruit growing regions around the world have further emphasized the importance of developing automation and robotic solutions for various field operations. In the last two to three years, a number of research institutions (e.g. Washington State University) and private companies (e.g. FFRobotics, Israel) have been funded publicly or privately to expand research and development efforts in this area.

4.8 Conclusions

Based on what has been accomplished in the past and what is happening currently around the world, we conclude that:

(1) Accuracy, speed and robustness of perception, manipulation and end-effector technologies for orchard applications have been improved rapidly over the past few years with artificial intelligence technologies, ever decreasing costs and increasing computational power including parallel computing, and advanced sensing and soft-robotic technologies.

(2) Advance in component technologies and significant public and private funding, automated and robotic technologies for various orchard operations, particularly for fruit harvesting, have advanced to the point that commercial availability and adoption of these machines seems inevitable in the near future.

(3) Advances in robotic picking technologies have also motivated the industry and academia to look beyond and further advance robotic technologies for many other orchard operations including tree canopy training, pruning, thinning, spraying and pollination.

(4) To expedite the development of automated orchard technologies further with the desired speed, accuracy, robustness and cost, the authors believe that increased efforts should focus on researching efficient algorithms and methods for multi-machine cooperation and human–machine collaboration while embracing trans-disciplinary research approaches and multi-purpose machine designs.

References

Amatya S (2015) Detection of cherry tree branches and localization of shaking positions for automated sweet cherry harvesting. PhD Dissertation, Washington State University

Baeten J, Donn K, Boedrij S, Beckers W, Claesen E (2007) Autonomous fruit picking machine: A robotic apple harvester. Springer Tracks Adv Robot 42:531–539

Baeten J, Donne K, Boedrij S, Beckers W, Claesen E (2008) Autonomous fruit picking machine: A robotic apple harvester. Field Serv Robot 42:531–539

Barth R, Hemming J, van Henten EJ (2016) Design of an eye-in-hand sensing and servo control framework for harvesting robotics in dense vegetation? J Biosyst Eng. https://doi.org/10.1016/j.biosystemseng.2015.12.001

Bulanon DM, Kataoka T (2010) Fruit detection system and an end effector for robotic harvesting of Fuji apples. Agric Eng Int: CIGR J 12(1):203–210

Beder C, Bartczak B, Koch R (2007) A comparison of PMD-cameras and stereo-vision for the task of surface reconstruction using patchlets. In: 2007 IEEE conference on computer vision and pattern recognition. IEEE, pp 1–8

Davidson JA (2016) Mechanical design and field evaluation of a robotic apple harvester. Ph.D Dissertation, Washington State University

Davidson J, Silwal A, Karkee M, Mo C, Zhang Q (2016) Hand-picking dynamic analysis for undersensed robotic apple harvesting. Trans ASABE 59(4):745–758

D'Esnon A, Rabatel G, Pellenc R, Journeau A, Aldon MJ (1987) MAGALI: a self-propelled robot to pick apples. ASAE Paper No. 87-1037. St. Joseph, MI

De Kleine ME, Karkee M (2015) Evaluating a non-Newtonian shear thickening surface during fruit impacts. Trans ASABE 58(3):907–915

Diener RG, Mohsenin NN, Jenks BL (1965) Vibration characteristics of trellis-trained apple trees with reference to fruit detachment. Trans ASAE 8(1):20–24

Elfiky NM, Akbar SA, Sun J, Park J, Kak A (2015) Automation of dormant pruning in specialty crop production: An adaptive framework for automatic reconstruction and modeling of apple trees. In: 2015 IEEE conference on computer vision and pattern recognition workshops (CVPRW). IEEE, pp 65–73

Eizicovits D, Berman S (2014) Efficient sensory-grounded grasp pose quality mapping for gripper design and online grasp planning. Robot Auton Syst 62(8):1208–1219

Emery KG, Faubion DM, Walsh CS, Tao Y (2010) Development of 3-D range imaging system to scan peach branches for selective robotic blossom thinning. ASABE, Paper No, p 1009202

Fernández JE, Cuevas MV (2010) Irrigation scheduling from stem diameter variations: a review. Agric Forest Meteorol 150(2):135–151

Galinato S, Gallardo RK (2011) 2010 Estimated Cost of Producing Pears in North Central Washington (FS031E). http://extecon.wsu.edu/pages/Enterprise_BudgetsAccessed 7 Jan 2013

Gallardo RK, Brady MP (2015) Adoption of labor–enhancing technologies by specialty crop producers: the case of the Washington apple industry. Agric Finan Rev 75(4):514–532

Gao M, Lu TF (2006) Image processing and analysis for autonomous grapevine pruning. In: 2006 International conference on mechatronics and automation. IEEE, pp 922–927

Glover JD, Reganold JP, Bell LW, Borevitz J, Brummer EC, Buckler ES, Xu Y (2010) Increased food and ecosystem security via perennial grains. Science 328(5986):1638

Gongal A, Amatya S, Karkee M, Zhang Q, Lewis K (2015) Sensors and systems for fruit detection and localization: a review. Comput Electron Agric 116:8–19

Gonzalez-Barrera A (2015) More Mexicans leaving than coming to the U.S. Pew Research Center. http://www.pewhispanic.org/files/2015/11/2015-11-19_mexicanimmigration__FINAL.pdf. Accessed 24 Nov 2015

Harrell RC, Adsit PD, Munilla RD, Slaughter DC (1990) Robotic picking of citrus. Robotica 8(1990):269–278

Hayashi S, Shigematsu K, Yamamoto S, Kobayashi K, Kohno Y, Kamata J, Kurita M (2010) Evaluation of a strawberry-harvesting robot in a field test. Biosyst Eng 105:160–171

He L, Zhou J, Du X, Chen D, Zhang Q, Karkee M (2013) Energy efficacy analysis of a mechanical shaker in sweet cherry harvesting. Biosyst Eng 116(4):309–315

Hedden SL, Whitney JD, Churchill DB (1984) Trunk shaker removal of oranges. Trans Am Soc Agric Eng 27:372–374

Hood CE, Mchugh CM, Sims ET, Garrett TR, Williamson RE (1981) Orchard fruit handling system. Trans ASAE 24(1):20-0022

Karkee M, Steward BL, Tang L, Abd Aziz S (2009) Quantifying sub-pixel signature of paddy rice field using neural network. Comput Electron Agric 65(2009):65–76

Karkee M, Zhang Q (2012) Mechanization and automation technologies in specialty crop production. Trans ASABE 19(5):16–17

Karkee M, Steward B, Kruckeberg J (2013) Automation of pesticide application systems. In: Zhang Q, Pierce F (eds) Agricultural automation: fundamentals and practices. ISBN: 9781439880579. CRC Press, Boca Raton, Florida, USA

Karkee M, Adhikari B, Amatya S, Zhang Q (2014) Identification of pruning branches in tall spindle apple trees for automated pruning. Comput Electron Agric 103(2014):127–135

Karkee M, Adhikari B (2015) A method for three-dimensional reconstruction of apple trees for automated pruning. Trans ASABE 58(3):565–574

Karkee M, Zhang Q, Whiting M, He L, Fu H, Xia H (2016) Localized shake-and-catch harvesting for fresh market apples. In: CIGR-AgEng Conference, June 26–29, Aarhus, Denmark

Karkee M, Silwal A, Davidson JR (2018) Mechanical harvest and in-field handling of tree fruit crops. In: Editor: Qin Z (ed) Automation in tree fruit production, principles and practice

Kondo N, Shibano Y, Mohri K, Monta M (1994) Basic studies on robot to work in vineyard 2: discriminating, position detecting and harvesting experiments using a visual sensor. J Japanese Soc Agric Mach 56(1):45–53

Lang GA, Whiting M, Long LE, Musacchi S (2015) Cherry training systems. https://research.lib raries.wsu.edu:8443/xmlui/handle/2376/6087. Accessed 20 Aug 2016

Lee MF, Gunkel WW, Throop JA (1994) A digital regulator and tracking controller design for an electrohydraulic robotic grape pruner. In: Proceedings of the 5th international conference on computers in agriculture, pp 23–28. ASAE, St. Joseph, MI

Lyons D, Heinemann P (2016) Selective automated blossom thinning. U.S. Patent 20,160,057,940

Ma S, Karkee M, Fu H, Sun D, Zhang Q (2016) Air suspension-based catching mechanism for mechanical harvesting of apples. In: Proceedings of the 5th IFAC conference on sensing, control and automation technologies for agriculture, Seattle, WA, 14–17 Aug 2016

Monta M, Kondo N, Shibano Y (1995) Agricultural robot in grape production system. In: International conference on robotics and automation, Nagoya, Japan, pp 2504–2509

Monta, M., Kondo, N., & Ting, K. C. (1998). End-Effectors for Tomato Harvesting Robot. Artif Intell Rev 12:11–25

Muscato G, Prestifilippo M, Abbate N, Rizzuto I (2005) A prototype of an orange picking robot: past history, the new robot and experimental results. Ind Robot: Int J 32(2):128–138

Naugle JA, Rehkugler GE, Throop JA (1989) Grapevine cordon following using digital image processing. Trans ASAE 32(1):309–0315

Ochs ES, Gunkel WW (1993) Robotic grape pruner field performance simulation. Paper no. 933528. ASAE, St. Joseph, MI

Ortiz C, Blasco J, Balasch S, Torregrosa A (2011) Shock absorbing surfaces for collecting fruit during the mechanical harvesting of citrus. Biosyst Eng 110:2–9

Pacheco A, Rehkugler GE (1980) Design and development of a spring activated impact shaker for apple harvesting. Trans ASAE

Parchomchuk P, Meheriuk M (1996) Orchard cooling with pulsed overtree irrigation to prevent solar injury and improve fruit quality of Jonagold'Apples. HortScience 31(5):802–804

Parameswarakumar M, Gupta CP (1991) Design parameters for vibratory Mango harvesting system. Trans ASAE 34(1):14–20

Peterson DL, Wolford SD (2003) Fresh–market quality tree fruit harvester Part II: apples. Appl Eng Agric 19(5):545

Polat R, Gezer I, Guner M, Dursun E, Erdogan D, Bilim HC (2006) Mechanical harvesting of pistachio nuts. J Food Eng 79(4):1131–1135

Rabatel G, Bourely A, Sevila F, Juste F (1995) Robotic harvesting of citrus: state-of-art and development of the French Spanish EUREKA Project

Sakai S, Iida M, Osuka K, Umeda M (2008) Design and control of a heavy material handling manipulator for agricultural robots. Auton Rob 25(3):189–204

Scarfe A, Flemmer R, Bakker H, Flemmer C (2009) Development of an autonomous kiwifruit picking robot. In: 4th international conference on autonomous robots and agents, Wellington, New Zealand

Seo HH (2003) Site selection criteria for the production of high quality apples based on agroclimatology in Korea. Kyung Hee University, 130 pp

Sevilla F (1985) A robot to Prune the Grapevine. AgriMation 1:190–199. ASAE, St. Joseph, MI

Sharda A, Karkee M, Zhang Q, Ewlanow I, Adameit U, Brunner J (2015) Effect of emitter type and mounting configuration on spray coverage for solid set canopy delivery system. Comput Electron Agric 112:184–192

Silwal A, Davidson J, Karkee M, Mo C, Zhang Q, Lewis K (2016) Design, integration and field evaluation of a robotic apple harvester. J Field Robot: (Under review: Accept after minor revision)

Sutyemez M (2011) Pollen quality, quantity and fruit set of some self-compatible and self-incompatible cherry cultivars with artificial pollination. Afr J Biotech 10(17):3380–3386

Tanigaki K, Fujiura T, Akase A, Imagawa J (2008) Cherry-harvesting robot. Comput Electron Agric 63:65–72

Umeda M, Kubota S, Iida M (1999) Development of "STORK", a watermelon-harvesting robot. Artif Life Robot 3:143–147

Thomas CS, Skinner PW, Fox AD, Greer CA, Gubler WD (2002) Utilization of GIS/GPS-based information technology in commercial crop decision making in California, Washington, Oregon, Idaho, and Arizona. J Nematol 34(3):200

United States Department of Agriculture Economic Research Service (2010) Labor-intensive US Fruit and Vegetable Industry Competes in a Global Market, Washington, DC, ERR-106, pp 57

Whiting M (2008) The UFO system: a novel architecture for high efficiency sweet cherry orchards. In: Proceeding of the IXth international symposium on integrating canopy rootstock and environmental physiology in orchard systems

Ye Y, Wang Z, Zhang Y, Jones D, He L, Hollinger GA, Taylor ME, Smart WD, Zhang Q (2017) Bin-dog: a self-propelled platform for bin management in orchards. Comput Electron Agric

Zhao D, Lu J, Ji W, Zhang Y, Chen Y (2011) Design and control of an apple harvesting robot. Biosyst Eng 110:112–122

Zhou J, He L, Zhang Q, Karkee M (2014) Effect of excitation position of a handheld shaker on fruit removal efficiency and damage in mechanical harvesting of sweet cherry. Biosyst Eng 125:36–44

Chapter 5
Robotics for Precision Viticulture

Francisco Rovira-Más and Verónica Saiz-Rubio

5.1 Technological Needs, Barriers, and Current Solutions for Competitive Vineyards

Grapes are included in what are called specialty crops, fruits of high return value typically set in orchards, which account for 50% of the total value of crop production in the USA, accounting for $60 billion in 2005 (Burks et al. 2008) and reaching $76 billion in 2012 (USDA 2012). In Europe, specialty crops are valued at about 70 billion Euro per year, representing 22% of the total output value of the agricultural sector in 2014. The fruit and vegetable sector alone accounts for about 45 billion Euro, with a total production of 40 million tons of fruit. In the transformation of grapes into wine, Europe is the global market leader accounting for 45% of the world's wine-growing area in 2014, 65% of production (167 Million hectolitres), 57% of global consumption and 70% of exports in global terms (Wine Institute 2016). The stable and privileged position of a wine in the global market depends on its long-term reputation, which takes considerable effort to attain but can be lost rapidly when a given standard is not assured. It is a known fact in viticulture that the **best wine is made in the vineyard rather than in the winery,** because grapes of high quality are the best guarantee for producing excellent wines. When the grapes are medium quality, efforts in the winery might correct certain defects, but will unlikely lead to premier wines. In Europe, Spain, France and Italy account for 32% of the total area devoted to vineyards in the world, followed by China with 11%, Turkey with 7% and the USA with 6% (Fig. 5.1a). The total production of wine in the world has increased 6.4% from 26,544 million litres in 2011 to 28,230 million litres in 2014. France, Italy, Germany and Spain alone account for 49% of global production, that is, almost half

F. Rovira-Más (✉) · V. Saiz-Rubio
Universitat Politècnica de València, Valencia, Spain
e-mail: frovira@upv.es

© Springer Nature Switzerland AG 2021
A. Bechar (ed.), *Innovation in Agricultural Robotics for Precision Agriculture*,
Progress in Precision Agriculture,
https://doi.org/10.1007/978-3-030-77036-5_5

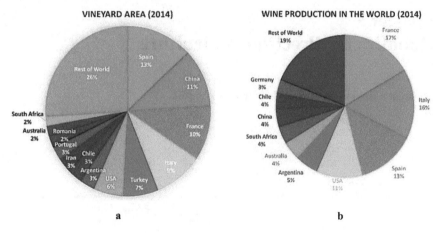

Fig. 5.1 World figures in the wine industry: **a** vineyard area and **b** wine production

of the worldwide production as shown in Fig. 5.1b (Wine Institute 2016), resulting in 13,833 million litres of wine being produced in these four European countries only.

It is possible to make great wine by chance or a good recipe, but not consistently; only by measuring and controlling key factors can the best wine be ensured year after year (Cox 1999). To make good wine consistently, it is necessary to test the grapes weekly to associate certain tastes with certain changes in measured properties. Weekly monitoring in the crucial weeks preceding harvest will allow the development of new management strategies for harvesting grapes in diverse zones at different times, which avoids mixing grapes of different degrees of maturity, a common source of poor wines. However, modern production of wine grapes, i.e. that based on objective and precise field data, is inefficient for the majority of growers for the following reasons:

- **Monitoring cost:** it is expensive to acquire field data. It can be done only once a year in general, which deters the updating of field information and of assessing the evolution of vines during the growing season. The nitrogen content, for example, varies continuously as fertilizer or water are applied.
- **Low rate of sampling:** it is not feasible to ask an operator to obtain sample data every metre, and as a result, measurements are usually sparse, say once every 400 m^2 (20 m × 20 m). With data from a sparse sample, conclusions can easily be biased. As a result, a weekly assessment of grape ripening in the six weeks prior to harvesting tends to be either too sparse or unaffordable.
- **Weight of current hand-held devices:** recording data for hours with a handheld device of several kilograms of mass (typically 2–6 kg) becomes exhausting for the operators, who also have to walk in the sun, usually in the summer.
- **Costs of service providers:** there are some service providers who can provide maps from airborne information, but they tend to be of low resolution. If several measurements are needed to determine how the plants evolve during the season,

the cost of around 80 € ha^{-1} (8camera 2016) means that monitoring 30 ha six times, for instance, will cost 14,400 € per season, which is prohibitive. The easiest way to have full control of information is by having full control of the scouting machine: you pay once and can use it as many times as needed. This is how farmers and field managers typically want to operate with machinery, and consequently is a promising approach to reach commercial success.

– There are **very few suitable commercial robots** to work in vineyards or other open agricultural fields. The majority of robots that exist today are at the laboratory stage and typically represent scientific proofs-of-concept. They are too complex and not reliable enough to cope with a 6- to 8-hour working timeframe. Some initial initiatives, however, are appearing but they only operate in small areas where technicians can assisst quickly and cost-efficiently.

5.1.1 Fertilization, Nutritional Status and the Estimation of Nitrogen Content

The *greenness* of a plant has traditionally been an accepted indicator of plant health. Some handheld devices such as SPAD® (Spectrum Technologies, Inc., Aurora, IL, USA) are used to determine deficiencies in leaves, mainly nitrogen, by estimating chlorophyll activity. However, these small meters have to be clamped over leafy tissue to calculate the chlorophyll index. Even though they are non-invasive, the need to clamp the leaves prevents these devices from being implemented on vehicles, and are consequently not efficient for robotic applications. An indirect way to assess nitrogen content, and therefore its deficiency, has been done by plotting the *normalized difference vegetation index* (NDVI) of canopies, an optical method based on the enhanced reflectance from a healthy canopy in the infrared spectrum. Differences in reflectance can be considerable between weak and healthy plants. The NDVI can be monitored from the air. An aerial map covering 10 ha with approximately 200 images obtained at a height of 80 m and reaching a resolution of 30 mm pixel^{-1} might cost around 800 € plus transport of equipment and operators to the test site (8camera 2016). On the other hand, NDVI can also be determined from a ground vehicle such as a conventional tractor, a utility vehicle or a robot. Some handheld devices can be fixed to conventional farm equipment to monitor nitrogen content (Fig. 5.2). Alternatively, machine vision in the near infrared band can be used to estimate the relative variation of a vine's canopy coverage within a vineyard, when images are recorded from the top of a moving vehicle equipped with a GPS to generate a map of plant vigour (Saiz-Rubio and Rovira-Más 2013).

Fig. 5.2 Ground-based NDVI estimation: **a** CropCircle® ACS-470 kit as a handheld device and **b** GreenSeeker® RT200C mounted on a conventional farm vehicle

5.1.2 Pruning and Pre-pruning

Pruning is a crucial operation in viticulture because it influences the development of the vine in the forthcoming season. Although pre-pruning the vines is easy to mechanize, pruning them in winter requires skills typically gained through years of experience, and it is consequently done manually by dexterous operators. Because of the lack of skilled labour in the winter to perform this operation, some wine-producing areas in Europe have indicated the need to introduce automation for this delicate task. A pruning robot was developed by Botterill et al. (2017). It is a mobile platform that straddles the row of vines. The plants are completely covered, such that sunlight is blocked to benefit computer vision processing. Images are taken with three cameras as it moves. The computer vision system builds a three-dimensional model of the vines and an artificial intelligence (AI) system decides which canes to prune. An articulated arm of six degrees of freedom executes the cuts.

5.1.3 Irrigation and the Control of Water Stress

Some vineyards, and even entire wine-producing areas, do not use irrigation in vine-yard management. However, when available, precise control of water stress by suitable rates of irrigation might become an influential factor in the final quality of the grapes and of the future wine. Such control requires constant feedback on the state of the plant, which evolves continuously during the production season and especially in relation to the weather. The measurement of canopy temperature as an indicator of stress was first identified in practice in 1981 (Jackson et al. 1981) with the definition of the Crop Water Stress Index (CWSI). Temperature differences between stressed and unstressed plants have encouraged the use of thermal images to assess water status. In addition, the continuous decrease in cost of compact thermographic cameras that can be mounted on agricultural vehicles, and even small aerial vehicles, has extended its use from initial defense applications to commercial civilian use including agriculture. However, there are still many technical hitches that limit their generalized use for automated solutions from field vehicles (Stoll and Jones 2007):

- The monitoring of stomatal activity in leaves requires the robust exclusion of sky, soil and grapes from infrared images.
- Sunlit canopies give a far wider range of temperature variation than shaded areas.
- Reference surfaces are required to calibrate the thermal images and estimate the temperature of leaves under wet and dry conditions prior to applying the CWSI.

5.1.4   Grape Harvesting: Deciding the Most Critical Moment for Winemaking

According to Cox (1999), there are three factors, sugar, titratable acidity (TA) and pH, that can be tracked weekly after *véraison* (colour change in red grape berries identifying ripening) and that will reach optimum levels when the grapes are ready to harvest. The pH is related to TA, but differs from it in significant ways because the pH of grape juice might or might not be correlated with the amount of tartaric acid. Unfortunately, these three properties require some grapes to be obtained and the juice extracted to measure these chemistry-based properties. This makes it impossible to measuring them 'on-the-fly' and by non-invasive techniques (fast measurements without touching the grapes), which are fundamental for an automated solution such as airborne imagery (remote sensing) and ground-based platforms carrying monitoring sensors onboard (proximal sensing from farm equipment and field robots). In general, the statistical significance of these measurements is weak because of the lack of intensive sampling. In addition, laboratory analyses require several days to obtain the data and are typically too costly for average producers if they want to have a well-sampled vineyard. The monitoring of traditional key properties that determine the ripening status of grapes, namely sugar, acidity and pH, has to be done manually by sampling certain bunches in the field. According to experts, it should be done regularly in the weeks before harvesting to obtain meaningful results. As the grapes grow under the canopy, aerial images cannot reach them, therefore remote sensing and proximal sensing from aerial images (unmanned aerial vehicles or *drones*) cannot be used for this purpose. Only ground-based monitoring is feasible for determining the maturity of red grapes.

The measurement of *anthocyanins* in the (red) grape skin provides an alternative and indirect method to assess maturity. Anthocyanins have been chosen as markers of phenolic maturation because their evolution with ripening is equivalent to that of skin tannins (Agati et al. 2007). This has resulted in the development of new sensors. The spectrophotometer Spectron® (Fig. 5.3a) announced by Pellenc (Matese and di Genaro 2015) and the Multiplex® (Fig. 5.3b), released by Force-A (Orsay Cedex, France), are two examples of the growing interest in developing handheld sensors. However, there is currently no off-the-shelf sensor that can estimate the maturity status of grapes from a moving platform before harvesting. To obtain a map of

Fig. 5.3 Handheld maturity assessment: **a** Spectron® (Pellenc) and **b** Multiplex® (Force-A)

anthocyanins before harvesting and at an adequate sampling rate would require the services of a company with a handheld device to walk along the rows and make multiple measurements. The data, the map and its scientific interpretation would also incur costs.

The generation of a manual map of anthocyanin levels in red grapes is possible with a handheld device such as those depicted in Fig. 5.3. Typical coverage might involve measuring 400 bunches per hectare, i.e. a point every 10 m × 10 m as the average of four representative grape bunches. At present, there is no commercial device to measure the anthocyanin content on-the-fly. Even though the European-funded VineRobot project (VineRobot 2014) worked for three years on a novel device to assess anthocyanins from a moving robot by combining computer vision and fluorescence, it was not feasible to measure anthocyanin levels at a distance of 40 cm from the grapes. Fluorescence-based sensors like the one shown in Fig. 5.3b usually analyse a circular spot of reduced diameter, typically between 4 and 8 cm. If the spot contains over 3% of green matter, the fluorescence reading is usually unreliable, and therefore must be discarded from maps of maturity. The reason behind this rationale is the large response of green matter to fluorescence when compared to anthocyanins, which typically masks the readings even with few leaves, stems, tendrils or peduncles. The resolution affordable with handheld devices, say 400 measurements per ha, would result in working cells of 100 m^2. However, an automatic system onboard a robot could obtain several measurements per metre in each scanned row if grapes are not overly occluded. This can be prevented by defoliation, a traditional operation in many vineyards to increase the sun's radiation on maturing bunches.

5.2 Semi-autonomous Solutions: Decision-Making for Man-Driven Vehicles

A middle ground on the way to autonomous solutions for vineyard management might be affordable to many growers who already possess state-of-the-art farm machinery.

The unprecedented availability of massive data sets from a new batch of novel sensors, as described in the previous section, can lead to a new way to use standard machinery. Robust machines that have proved reliable in the field, will increase their efficiency when intelligent decisions resulting from data recorded precisely in the field are added to the decision-making process. The following examples explain how to use nitrogen content in leaves and the amount of anthocyanin in grapes to enhance the performance of fertilizers and grape harvesters.

5.2.1 Variable-Rate Fertilization with Prescription Maps

The determination of vegetation indices from moving platforms, as shown in Fig. 5.2b, can provide a basis for variable-rate application of fertilizers in the vineyard so that vigorous vines do not get an excess of nutrients and weaker plants receive what they need to produce a satisfactory yield. The rate of fertilizer application can be determined from a digital prescription map that the machine understands. Such a map will include spatial information for locating the vehicle in the field and the recommended dose to apply. A highly varying dose as the vehicle moves is not usually practical, but if homogeneous zones with similar needs are identified, different rates can be applied efficiently to specific areas of the field, provided the data have been mapped accurately. These zones might be large areas, or alternatively, square cells with sides ranging from 1 to 10 metres according to the accuracy of the sensor, the resolution of the map and the grower's management approach. Figure 5.4a shows a field map depicting the *nitrogen balance index* (NBI; Cerovic et al. 2015), an indirect way to assess the variation in foliar nitrogen in a vineyard recorded by a robot along the trajectory of Fig. 5.4b, which was registered with an onboard GPS receiver. The rows are spaced 2.4 m apart and have a length of approximately 105 m. The

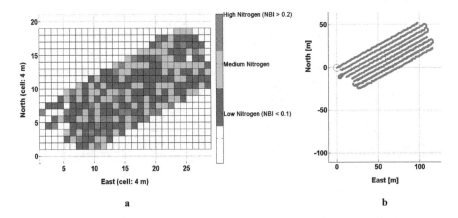

Fig. 5.4 Real-time NBI maps: **a** grid maps with 4 m × 4 m cells and **b** vehicle trajectory

cells represented in Fig. 5.4 are squares of 4-m side, which have been obtained by averaging all the measurements within the 16 m^2 of each cell.

The NBI is the ratio of chlorophyll content to epidermal flavonol leaf content, and can be used as an indicator of the nutritional status of the plant. It is less sensitive to phenology because it reflects the availability of nitrogen better than the two indicators used separately, which have inferior quality as estimators. The statistical correlation between NBI and nitrogen content in mg per gram of leaf, unfortunately depends on the side of the leaf being measured and the cultivar, but a practical relation for adaxial measurements on Pinot Noir vines, for example, was given by the equation NBI = −0.4 + 0.62 N (mg g^{-1}) (Vinerobot 2014). Even though space has been discretized in Fig. 5.4a to avoid intense changes in the actuation of the solenoids that adjust fertilizer doses, the prescription maps could be simplified further by reducing the doses to a smaller set of choices, such as *high* and *low*, for example. This downscaling of rates can be achieved by several approaches, from a simple resetting of rates to more sophisticated clustering techniques. Kriging (Oliver 2010) has been used in precision agriculture to smooth data that vary spatially and to interpolate from relatively sparse data that is spatially correlated. Kriging involves predicting values from neighbouring data at unsampled places using the model parameters fitted to an experimental variogram, therefore, it requires sufficient data from which to compute a variogram; at least 100 points.

Figure 5.5a shows a grid map with 4-m square cells filled with nitrogen measurements from a fluorescence sensor (NBI*100), and all the measurements within a given cell were averaged. The median value of the 25 cells considered for each moving window was applied to each averaged cell (Saiz-Rubio and Rovira-Más 2016) to obtain the simplified map of Fig. 5.5b, which would be more practical for field operations. This new map has resulted in zones with similar characteristics,

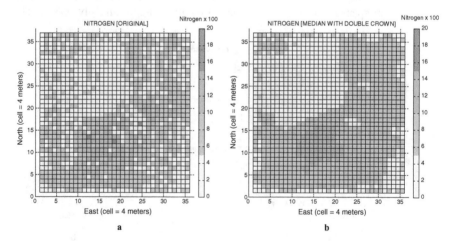

Fig. 5.5 Smoothing operations in field maps as a basis for variable-rate applications: **a** raw measurements and **b** smoothed map with moving averages

facilitating the future implementation of automated tasks such as variable-rate applications (VRA) of fertilizer. Figure 5.5b can be used as the basis for a prescription
map to fertilize according to nitrogen deficiencies detected in the field and recorded
by a robotic platform. Agricultural geographic information systems (GIS) software
is used to create prescription maps. A prescription map tells the controller of an
intelligent vehicle how much product has to be applied at each location of the field.
Most agricultural GIS packages can create prescription maps in multiple formats
(Norwood et al. 2009). Further research, however, will be needed to determine if
these procedures have any effect on crop growth and fruit bearing, which is the final
objective of applying precision techniques in the vineyard.

5.2.2 Differential Harvesting with Intelligent Mechanical Harvesters

A long-term wish of wine makers and vineyard growers has been differential
harvesting, in which a field is harvested at different periods to avoid mixing grapes
of uneven maturity. Until now, this has not been practical either for most manual
harvesting or with mechanical harvesters. However, the advent of new machines
that can read GPS instructions and interpret digital maps provides the potential for
differential harvesting, especially with vehicles that can carry two independent bins
where grapes may be placed according to onboard computer commands. Although
cutting-edge harvesters endowed with intelligent behaviour and new physical capabilities will be necessary for advanced harvesting techniques, there are still important
steps that need to be solved before differential harvesting can be achieved, such as
the provision of precise harvest-readiness maps. The anthocyanin level of red grapes
will be an important component in such maps, but other complementary properties
might help, such as the nitrogen content in leaves. Figure 5.6 shows a plot of the
evaluation of four wines by scoring their main oenological properties on a 0–5 scale.
The four wines come from the same vineyard, but the grapes used to make them
come from separate sub-zones with distinct contents of nitrogen in the leaves (N)
and anthocyanins in the grapes (A). Two levels for each property were established
(*high* and *low*), resulting in four combinations A + N–, A–N–, A + N + and A–N +
. Wine tasting experts finally concluded that the best wine was that made with a large
anthocyanin content and a small nitrogen content (A + N–), as plotted in Fig. 5.6.

The mathematical combination of several maps, each one plotting a relevant field
property, is feasible provided their axes, coordinates, origin and units are compatible.
The grid maps of Figs. 5.4 and 5.5, with a local origin and plain coordinates East
and North, provide a convenient way to fuse field data. The nitrogen distribution of
Fig. 5.4a could be further simplified to only two levels (N + and N–) and used to
group harvesting zones by following the philosophy of Fig. 5.6. Based on a cell-
to-cell comparison, maps generated automatically from a moving vehicle could be
fused with manually-generated maps such as those showing the spatial distribution

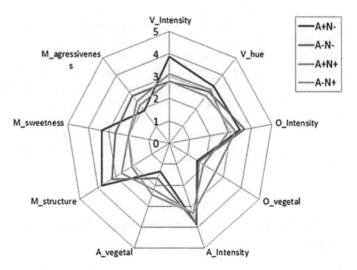

Fig. 5.6 Wine properties (**V**isual; **O**dour; **A**roma; **M**outh) as a combination of fundamental field properties nitrogen (**N**) and anthocyanins (**A**) (Courtesy of Les Vignerons de Buzet, France)

of titratable acidity, must pH, sugar content or yield, to define a quality index for the future wine (Rovira-Más and Saiz-Rubio 2013). The various properties of the wines represented in Fig. 5.6 indicate that, as expected, the amount of anthocyanin in red grapes is an important property for classifying the oenological potential of a wine. However, the discussion raised in Sect. 5.1.4 demonstrates that, even though manual measurements with handheld devices are feasible, for a sampling rate to provide statistical significance it is better to make measurements on-the-fly from a moving vehicle. Figure 5.7a depicts real-time generated maps of anthocyanins measured in a vineyard of Merlot grapes with an experimental fluorescence device carried by a vineyard robot. Notice that this map is less populated than the nitrogen map of Fig. 5.4a because of the need to measure anthocyanins at a moving spot in which reliable estimates only occur with less than 3% of green matter at the spot, the rest being occupied by the grapes. The actual trajectory followed by the robot is shown in Fig. 5.7b.

Interpolation techniques, such as kriging, have been extensively used to interpolate data at places where there are no measurements. This procedure, however, does not guarantee a better representation of reality than a map with empty spaces like Fig. 5.7a. In fact, smooth interpolated maps might mask very variable data whose averaging might lead to the wrong decisions being made, so caution must be the rule when analysing data with large dispersion. If advanced harvesters can currently carry two bins (A–B) at most, and the anthocyanins map is used to determine automatically into which bin grapes must be loaded, it makes no sense to produce digital maps of more than two levels. Figure 5.8 shows a simplification of the anthocyanins map of Fig. 5.7a to only two levels (*high* and *low*) to obtain a reasonable number of zones for applications similar to that illustrated in Fig. 5.6.

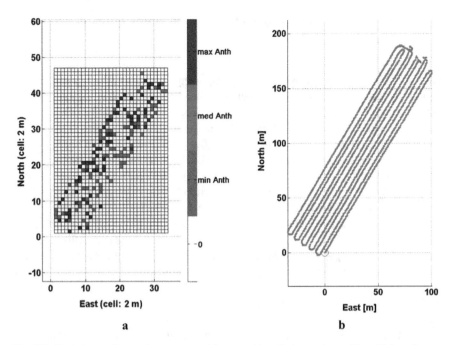

Fig. 5.7 Real-time anthocyanins maps: **a** grid maps with cells 4 m × 4 m and **b** vehicle trajectory

Fig. 5.8 Simplification of Fig. 5.7a for a more practical zoning with two levels of anthocyanins (±)

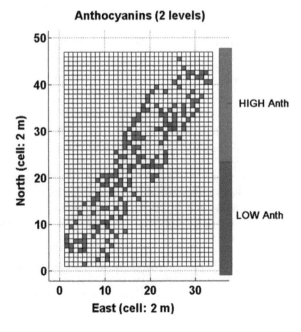

5.3 Autonomous Solutions: The Advent of Agricultural Robots

Even though most vineyards have vine-supporting structures that can assist navigation, the problem of autonomous guidance is a complex problem because of the considerable uncertainty and the extraordinary risks involved with farm machinery. A vehicle in the open field is subjected to many disturbances caused by a dynamic environment such as changing illumination, fluctuating weather and unpredictable obstacles including tools, other machines, animals, or even people working in the field. Barren fields ready to be sowed require guidance commands from satellite-based positioning systems for automating farming tasks; whereas, vineyards typically have vines following a particular arrangement. Robotics and automation greatly benefit from vertically-oriented supporting systems, such as VSP (vertical shoot positioning), rather than the more traditional goblet training system. Although agricultural robotics is growing at present, the commercial offer of farm robots is very limited, yet many research groups at universities, government agencies and private corporations are making considerable efforts to develop robotic solutions to actual problems found in agricultural fields. The following points address several crucial challenges in the long journey from basic semi-autonomy to fully autonomous farm robots. Initial attempts, as the platform of the products mentioned in Agati et al. (2007), give a good idea of the growing interest in these technologies.

5.3.1 Reliability and Safeguarding as the Highest Priority

The systematic accumulation of sensors in automated applications, not always indispensable, have often resulted in weak solutions when challenged by the harsh environments of farm fields over an extended period of time. There is a big difference between a 10-minute demonstration and regular equipment operations during the entire season. The trade-off between complexity and reliability is key, and as a result we should verify carefully that adding a new component is strictly necessary to meet the end-user requirements because each new component will involve more complexity, and therefore a greater likelihood of failure (Vinerobot 2014). Fail-safe conditions may be enhanced by introducing redundancy in the system and by designing a reliable safeguarding network. To do so, the following features should be considered:

- Robots are usually designed to be proficient in defined environments, thus no operations outside pre-defined settings should be allowed. In the case of viticulture, for example, robots should not operate outside the vineyards. Global navigation satellite systems (GNSS) receivers should warn when a vehicle leaves the confidence zone set by the user.
- Canopy or terrain disturbances may induce unstable behaviour in the navigation engine of robots, putting them at risk after getting too close to surrounding vines

or supporting structures. For such situations, it is necessary to stop the robot automatically and safely before it collides with other objects, and in case the non-contact system fails, halt the robot as soon as an obstacle is touched. For the latter case, a frontal bumper often becomes an efficient solution.

- There are many causes, some of them unpredictable, that can make a robot perform erratically or unstably; therefore, a network of emergency stop push buttons must be mounted and evenly distributed on the robot's exterior so that anyone in the vicinity can stop it without potential harm.
- An intelligent vehicle that can operate autonomously receives instructions from one or several computing units. If a power shortage affects the normal performance of the main computer and ancillary components, the consequences may be lethal for the robot's integrity. This is especially important when the robot is powered by electric batteries because the electronic network of the robot might behave randomly if battery power decreases below a threshold. Therefore, close monitoring of the power system is important for stability during the robot's operational time.

Figure 5.9 shows the safety network implemented in the first prototype of the VineRobot (2014). Four emergency stop (E-stop) push buttons have been placed near each corner of the four-sided body of the robot (only two are visible in Fig. 5.9). When any of the buttons is activated, a relay cuts the power to the wheel motors and turns on the red warning light at the same time the buzzer sounds. The three sonar sensors mounted on the bumper and facing forward are programmed to stop

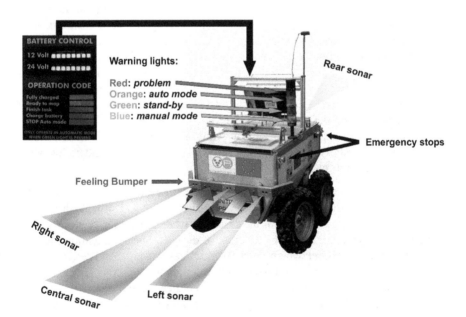

Fig. 5.9 Explanatory diagram of a safety network for a vineyard robot

the robot if an obstacle is detected at less than 50 cm from its front. Similarly, the rear sonar provides assistance for reverse manoeuvres at the headlands. If the frontal ultrasonic sensors do not halt the robot before an immediate collision, a gentle push on the bumper would fire the same relay that stops the wheel motors and issues an acoustic warning. When the robot of Fig. 5.9 was evaluated in actual vineyards, and after several hours of continuous operations in the field, a weak voltage in the battery system resulted in irregular behaviour of the stereo camera, which in turn froze auto-guidance images and eventually made the robot go astray. To avoid power-induced instabilities, the status of both the 12 VDC and 24 VDC power systems was tracked independently by a light bar display near the monitor, and also by an indicator included in the graphical user interface (GUI, yellow bars; Fig. 5.15a). When the voltage from either power system dropped below a predefined threshold, the robot sent a warning message and was stopped safely, disengaging automatic mode (orange light off) and only allowing manual operations (blue light on).

5.3.2 Physical Requirements and Mechanical Design

Field testing with robots in real environments has shown that it is important, especially in agriculture, not to overlook the mechanical design to focus only on sensors, electronics and software development. A robotic platform that is supposed to compete, and optimally outperform, conventional farm machines will have to traverse all kinds of uneven and rough terrain, perform many hours of continuous operation and endure tough outdoor conditions including unexpected rain, high humidity, extreme heat in the summer or cold in the winter, and occasional strong winds. Consequently, the mechanical structure of a robot must withstand friction, vibration, wear, vertical accelerations (shocks) caused by bumpy terrain and even occasional branches hitting or scratching its external cover. In addition, the power delivered by the batteries or combustion engine must be conveyed efficiently to the tyres, which implies making the right choice when designing the transmission system and the steering mechanism. Trying to solve mechanical problems with software tends to be futile and often catastrophic. The following list reviews some key aspects under consideration when designing agricultural robots:

- The materials with which the supporting frame and the external cover of the robot are built must be resistant to corrosion, waterproof and strong. Aluminium and steel are good candidates for the structure, whereas external bodies made of malleable polymers leave room for creative designs. Special attention must be paid to the joints through which water and dust may penetrate and deteriorate the inner electronics, typically not well fitted for outdoor conditions. The design of chassis and body must give priority to practical needs rather than aesthetics so that replacing a battery or repairing a linkage does not require dismantling the entire robot.

- Effective transmission involves selecting the right set of mechanical components so that the final torque and rotational speed in the tyres optimizes the available power for versatile performance. In general, moderately-sized robots powered by electric batteries cannot handle complex drivetrains comprising clutches, multi-gear boxes or torque converters. Rather, they benefit from simplified approaches in which a reduced set of gears link electric motors and tyres. Yet, the selection of these gears is crucial to ensure the expected performance of the robot in all foreseen situations; the wrong speed will make the robot inefficiently slow or dangerously fast, whereas a lack of torque will compromise its roving capacity.
- Regardless of the precision achieved in the navigation control commands, if they are not properly executed by the steering system, the robot will not reach the desired position at the right time. Therefore, it is essential to define the steering strategy and design of the steering mechanism. Sharp and small corrections are needed for straight guidance, but large wheel angles will be necessary to complete headland turns successfully. For Ackerman geometry, the wheelbase and maximum turning angle of the front wheels are critical parameters to determine the turning radius of the robot, which is key in the automatic execution of headland turns. An efficient way of protecting the steering actuators of autonomous vehicles, particularly electric motors, is by limiting the sweeping movement of tie rods with end-of-stroke switches, avoiding extreme angles, friction wear and overheating of drive cards (Rovira-Más et al. 2015a).
- It is impossible to predict the properties of the terrain where the future robot will have to navigate during its lifespan. Even if the terrains were known, the effect of weather and farming tasks on the ground would alter their tractional capacity. Consequently, a compliant suspension system can considerably improve the mobility of the robot in the vineyard by increasing the likelihood of keeping the four wheels in contact with the ground all the time. Wheel slippage is unavoidable in off-road terrain, but limiting it will have a positive effect on the navigational capabilities of robots, mainly when negotiating the sharp turns at the end of the rows to change the direction of travel 180° (Saiz-Rubio et al. 2017).
- Finally, the interior space storing the electronic components and computing units must be cooled efficiently to avoid processing slowness from overheating. Many agricultural tasks take place in the summer when ambient temperatures are high and the sun's radiation intense. The right location and choice of fans and ventilation grilles may effectively diminish the inside temperature of a robot and provide a safer environment for computers.

Figure 5.10 provides some examples of the mechanical components discussed above, such as a suspension system (a), a steering mechanism (b), cooling fans for the central computer of the VineRobot-II (c) and the open design of a robot that favours maintenance and assembly of new components (d) (Saiz-Rubio et al. 2017).

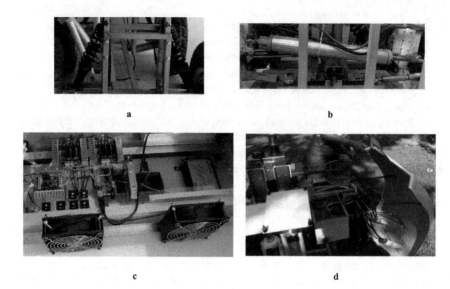

Fig. 5.10 Mechanical choices for agricultural robots: **a** suspension, **b** steering, **c** cooling and **d** frame

5.3.3 Fundamental Abilities: Navigation and Mapping

One of the most delicate and complex tasks entrusted to a robot is autonomous navigation. An operation that humans resolve effortlessly from childhood becomes a serious challenge for a machine when uncertainty is brought into the equation, as is the case in open agricultural fields and environments. An effective approach to cope with this challenge in row-structured agriculture is by dividing the auto-steering operation into two distinct stages: navigation between rows in a quasi-straight guidance, and headland turning to change rows after making a U-turn following a specific turning geometry. Figure 5.11 depicts both cases in a vineyard.

While global positioning by GNSS technology is vital for field mapping and precision farming applications, autonomous guidance in orchards and vineyards cannot rely on satellite-based navigation exclusively because precise steering commands cannot be ensured with signal blocking from trees or multipath errors induced by nearby structures or vines. As a result, local perception provides the complementarity needed to ensure a richer understanding of a robot's surrounding. Such perception is typically acquired by ultrasonic sensors, lidar rangefinders or any form of machine vision. Very often, the combination of various sensors, rather than just one, provides the level of accuracy required to guide a vehicle inside the tight space between adjacent rows, as shown in Fig. 5.11 (Buzet-sur-Baïse, France). Lidar and sonar have been extensively used to detect obstacles around a vehicle, but very often the guidance performance closest to human driving has been achieved with machine vision. When a camera is placed at the front of a vehicle, images with a vanishing point may be processed to find the optimal trajectory between crop rows, like a

a

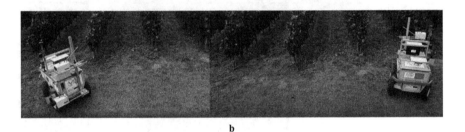

b

Fig. 5.11 Autonomous guidance of a vineyard robot: **a** inside-row guidance and **b** headland turning

monocular camera coupled with a near infrared filter in Rovira-Más et al. (2005), which used the Hough transform to determine the central path. A serious disadvantage of monocular cameras for outdoor conditions is their strong dependence on changes in ambient illumination, which often results in lack of robustness if conditions differ markedly during the operational time, typically ranging from dawn to dusk. Stereoscopic vision, however, can circumvent this shortcoming because two identical lenses, mimicking human eyes, perceive a scene by comparing the relative position of the same features in two imaging sensors, therefore changes in illumination affect both sensors simultaneously in such a way that as long as there is enough light intensity to find textural changes, pixels will be correlated and their coordinates estimated. Furthermore, the resolution of stereo geometry gives the three coordinates of a point in space, i.e. the three-dimensional (3-D) representation of the scene ahead of the robot, which represents a description of reality richer than the information contained in two-dimensional (2-D) images acquired with monocular cameras. Figure 5.12 shows the navigation maps derived from various situations perceived with a compact off-the-shelf stereoscopic camera. A multiplicity of algorithms may be applied to these navigation maps to find the steering command that will guide the robot along the vineyard rows. A particular example of image processing and its associated control system for stereo-based 3-D perception in autonomous navigation can be found in Rovira-Más et al. (2015a).

Fig. 5.12 Automatic guidance between vineyard rows with stereoscopic vision: navigation maps

Even though the majority of working time occurs travelling along the rows, which is where information is retrieved from plants or soil and agricultural actions must be executed, turning at the headlands to change rows is necessary for a continuous operation without human intervention. Consequently, it will be necessary to develop and encode a reliable algorithm to engage one row after another with agility, which is not a trivial endeavour. To begin with, the guidance features provided by bounding rows in straight guidance will no longer be available. To make things worse, slippage increases in sharp turns, and a slight deviation when entering the following row might result in unfortunate collisions. For all these reasons, this is a delicate manoeuvre that necessitates a special formulation. The row spacing, for example, will have an effect on the geometry of the turn, not to mention the special cases of rows of variable length to fit irregular fields, boundary rows near roads with traffic, or uneven headlands in sloping terrain.

A practical approach to deal with the headland turn problem has been by dividing the turning sequence into a set of consecutive stages where different sensing technologies are fused in such a way that each stage is solved with the best information available in the vehicle (Subramanian and Burks 2007). The ultrasonic sensor network of Fig. 5.9 was used to enhance straight navigation and assist in headland operations by the robot of Fig. 5.11, with the additional assistance of two lateral sonars pointing at the canopy, one on the left side of the robot and the other on its right side, resulting in a total network of six encircling sonars. Figure 5.13 shows a schematic diagram of the six stages into which a complete turn was decomposed,

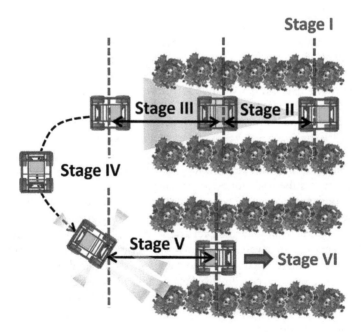

Fig. 5.13 Stages for executing headland turns by a vineyard robot

Table 5.1 Specifications for a multi-stage headland turning operation

Stage		Driving technique	Description
I	Detect end of row	3-D stereo vision	**Initiate turning mode** & quit straight guidance; reduce speed
II	Finish row with visual cues	3-D stereo vision	Use navigation map for guidance while the camera still perceives
III	Get out of row	Dead-reckoning	Advance the last 2–3 m to exit the row; use side sonars as perceptive information
IV	Turn 180°	Dead-reckoning	Steer to maximum angle ($\approx$ 20°) and straighten up
V	Transition stage	Sonar + 3-D stereo vision	Reduce speed & very slowly find the centre line; when necessary back-up (sonar fires) and re-try entry in the next row
VI	Engagement into new row	Sonar + 3-D stereo vision	If both side sonars and camera report a stable situation and there are no obstacles in front of bumper, **quit turning mode**

and graphically depicts the work of the six-sonar network at the end of stage IV. As explained in Table 5.1, each turn involved the combination of stereo vision, sonar and dead-reckoning to achieve a turn every two rows. Further details on these operations can be found in Rovira-Más et al. (2016).

Regardless of the navigation strategy chosen for an autonomous vehicle, a GNSS receiver will always provide valuable information for applications within the scope of robotics, precision farming and information technology (Rovira-Más et al. 2015b). The headland turning manoeuvre of Fig. 5.13, just to cite an example, uses GPS information to estimate the length travelled by the robot for stages III and IV that require dead-reckoning. In addition to navigation assistance, crop maps will benefit from global-based localization. However, the geodetic coordinates delivered by GNSS receivers through the NMEA code are not convenient for precision farming. Spherical coordinates such as latitude and longitude do not allow the use of Euclidean geometry, which is the basis for common calculations of distances and areas. The absence of a tangible origin of coordinates also complicates the creation and use of crop maps, whose final users are not typically experts in geographical systems. Earth sphericity can be neglected for relatively small areas such as vineyards, therefore UTM (universal transverse Mercator) or LTP (local tangent plane) coordinate systems are better adapted to robot-based mapping. The latter also allows end-users to choose the origin of the coordinate frame locally, what makes it ideal for users to correlate map zones within their own field. The LTP coordinate system, therefore, combines the advantages of global positioning with local coordinates East and North in a conventional Cartesian frame.

To make decisions based on objective data gathered from robotic platforms, as different sorts of data will be collected during the seasons, with diverse spatial resolution and measurement precision, a systematic way of correlating information in time and space will be necessary. An ordered division of field space into cells of meaningful size and agronomic significance allows the comparison of well-determined zones at a level of precision chosen by each user. However, the discretization of space into cells should not jeopardize the *global–local* advantages obtained with the LTP system. Fortunately, both approaches are compatible, and grids can be globally referenced in a Euclidean plane set to locate square cells by Cartesian coordinates East and North (Rovira-Más 2012). Moreover, this global-based grid approach allows for a real-time implementation as long as a GNSS receiver has been integrated properly in the mapping robot, as shown in Figs. 5.4 and 5.7. The raw data directly measured from the field by the onboard sensors are often too 'noisy' to make a map that can be read by growers or other machines. Geostatistics can be used to reduce the local noise in data reflected by marked changes over short distances (jumps). Based on the method of data processing chosen, maps will be available in real time, or alternatively, at the end of the mapping mission if the complete data set is needed to correct individual data points. In such cases, successive operations might be run immediately after mapping, leading to a *quasi-real-time* situation where maps are available from the field as soon as the robot has scanned the predefined area. An example following this approach is presented in Saiz-Rubio and Rovira-Más (2016).

5.3.4 Human–Robot Interaction in a User-Centred Design

Agricultural robots have to be designed with the premise that their future users are individuals used to handling tractors, harvesters, sprayers and other conventional equipment that is highly resistant, and straightforward to use and understand. Consequently, delicate, weak, highly-exposed, low-cost robots that work reasonably well indoors over firm and clean floors of research laboratories and unpolluted factories will never perform successfully and consistently in agricultural fields. Most agricultural robots are still at the research stages, and the complexity of handling and maintaining them is closer to experimental prototypes than commercial products. Efforts are currently being made to shrink this gap and make agricultural robots commercially available in less than a decade. The following paragraphs provide an overview of these secondary features that, without being central in the design of farm robots, are necessary to consider before deploying market-ready solutions.

The first point under discussion relates to transportation. A particular robotic solution may be integrated into a self-propelled platform such as a tractor or harvester, but in general a robot is designed to carry out a specific task in the field, and therefore must be carried from the storage building to the field and vice versa. This will restrict the size and weight of farm robots because the average user has to be capable of handling them without the need to purchase a new vehicle for this particular purpose. Conventional vans, utility vehicles, SUVs or pick-up trucks should suffice

Fig. 5.14 Preliminary steps in automated operations: transportation, unloading and placement of robots

for just one operator to move the robot from one field to the next. In addition to space requirements, users must be able to load and unload farm robots without making any physical effort greater than lifting a reasonable payload, 10 kg for example, which essentially forces the robots to be self-propelling in the loading operations. This can be facilitated by a joystick through which full control of the steering mechanism and wheel motion is provided. These joysticks can be linked to the robot wirelessly, but for such a delicate operation where tolerances may be limited and collisions are likely, wired remote controls provide a safer solution. Once the robot has been downloaded from the transporting vehicle, the joystick will allow the robot to be placed in the first row selected to begin automated tasks. Figure 5.14 illustrates the process of transportation, unloading and placement of a robot as the preliminary steps to carry out automated tasks.

After placing the robot in the first row, automated operations can begin. A straightforward and unambiguous interface should let users select the main features for each particular mission, facilitating the initiation of automated tasks. This interface will comprise hardware-based and software-based interactions. The former may include the power switches connected to batteries, warning lights indicating robot status (Fig. 5.9), or the button enabling automatic mode; the latter will be compactly outlined within a graphic user interface (GUI) manipulated through a touchscreen monitor integrated into the robot, and optionally remotely transferred to a mobile terminal. Figure 5.15a provides an example of a GUI for a vineyard mapping robot. Notice that, in general, this command window offers three types of information exchange between the robot and the user:

(a) *Visual information*: real-time video, battery level, 2-D navigation map and crop parameters cell map.
(b) *Textual information*: GPS data, row number and text messages.
(c) *Action buttons*: save data, velocity, mode (manual or auto), number of rows to map, etc.

As technology advances and new materials become popular and available, user preferences evolve with time. Modern farmers demand innovative solutions at the same technological level reached by other production sectors. The introduction of robotics in rural areas could encourage young farmers to modernize their equipment under the context of digital agriculture, as long as market demands allow for the economic sustainability of their investments. However, sustainability is also being considered nowadays from an environmental point of view. The implementation of renewable energy and recyclable materials are receiving more attention every day

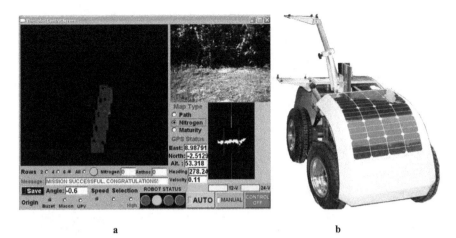

Fig. 5.15   VineRobot-II design features: **a** graphic user interface and **b** solar panels

among the manufacturers of agricultural equipment. Figure 5.15b shows a robot prototype with two plates of solar panels providing 60 W each (Saiz-Rubio et al. 2017).

5.4   Conclusions and Looking Beyond

A promising side effect of the successful introduction of robotics in commercial vineyards is the attraction that new technologies pose to young grape growers. The average age of farming populations in Europe and Japan is currently near retirement age, with very few professional farmers under 35 years old. The lure of electronics and automation will possibly help to counter the negative effect of an aging population in agriculture. This is one of the major problems faced by industrialized countries, especially with the potential demand for an increase of 100% in food with the growth in population expected in 2050. Figure 5.16b shows that there are many European farmers over 65 years old, which in many countries is the retirement age; and conversely, Fig. 5.16a shows the small number of farmers under 35 years old, the prototypical farmer who could give stability to the rural population in a *rural renaissance* induced by technology-based solutions.

In addition to the serious problem of an aging farming population, there are other issues for growers that make robotics attractive to viticulture. Among them, the shortage of skilled workers to prune vines in the winter, the lack of objective field data to maintain a wine of a certain quality consistently and its reputation over time, and the possibility of differential harvesting to avoid mixing grapes with different properties. Overall, there are many ways of improvement in viticulture through technology, but, on the other hand, there are also important challenges to overcome before reaching

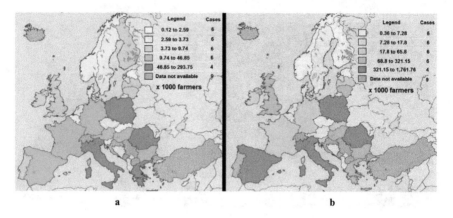

Fig. 5.16 European farmers under 35 years old (**a**) and over 65 years old **b** (Eurostat)

market readiness. The fact that robots are not widespread in vineyards worldwide suggests that these novel approaches remain difficult at the commercial and practical level. *Technical* challenges such as reliability and safety of autonomous vehicles operating several hours without human intervention, *economic* hurdles resulting from the need to use cutting-edge technology in products that must compete in price with other alternative solutions, and *social* barriers encountered when complex devices that produce unmanageable amounts of data have to be accepted and understood by an ever aging population all seem apparently insurmountable. However, recent progress in the fields of robotics and precision agriculture give much cause for optimism, and impressive innovations will soon reach the market and the global agronomic sector.

References

Agati G, Meyer S, Matteini P, Cerovic ZG (2007) Assessment of anthocyanins in grape (*Vitis vinifera* L.) berries using a non-invasive chlorophyll fluorescence method. J Agric Food Chem 55:1053–1061

Botterill T, Paulin S, Green R, Williams S, Lin J, Saxton V, Mills S, Chen X, Corbett-Davies S (2017) A robot system for pruning grape vines. J Field Robot 34:1100–1122

Burks TF, Schmoldt DL, Steiner JJ (2008) U. S. specialty crops at a crossroad. Resource (September). ASABE, St. Joseph, USA

Cerovic ZG, Ben Ghozlen N, Milhade C, Obert M, Debuisson S, Le Moigne M (2015) Nondestructive diagnostic test for nitrogen nutrition of grapevine (*Vitis vinifera L.*) based on Dualex leaf-clip measurements in the field. J Agric Food Chem 63:3669–3680

Cox J (1999) From vines to wines. Storey Publishing, North Adams (MA), USA

Jackson RD, Idso SB, Reginato RJ, Pinter PJ (1981) Canopy temperature as a crop water stress indicator. Water Resour Res 17(4):1133–1138

Matese A, di Genaro AF (2015) Technology in precision viticulture: a state of the art review. Int J Wine Res 7:69–81

Norwood SH, Winstead A, Fulton J (2009) Introduction to prescription maps for variable-rate Applications. Alabama Cooperative Extension System. ANR-1362. In: www.aces.edu. Accessed 17 Nov 2016

Oliver MA (ed) (2010) Geostatistical applications for Precision Agriculture. Springer, Dordrecht, Chapter 1, pp 12

Rovira-Más F, Zhang Q, Reid JF, Will JD (2005) Hough-transform-based vision algorithm for crop row detection of an automated agricultural vehicle. J Automob Eng 219(8):999–1010

Rovira-Más F (2012) Global-referenced navigation grids for off-road vehicles and environments. Robot Auton Syst 60:278–287

Rovira-Más F, Saiz-Rubio V (2013) Crop biometrics: the key to prediction. Sensors 13:12698–12743

Rovira-Más F, Millot C, Saiz-Rubio V (2015a) Navigation strategies for a vineyard robot. ASABE Paper n° 152189750. ASABE, St. Joseph, USA

Rovira-Más F, Chatterjee I, Saiz-Rubio V (2015b) The role of GNSS in the navigation strategies of cost-effective agricultural robots. Comput Electron Agric 112:172–183

Rovira-Más F, Saiz-Rubio V, Millot C (2016) Sonar-based aid for the execution of headland turns by a vineyard robot. ASABE Paper n° 162456431. ASABE, St. Joseph, USA

Saiz-Rubio V, Rovira-Más F (2013) Proximal sensing mapping method to generate field maps in vineyards. Agric Eng Int: CIGR J 15:47–59

Saiz-Rubio V, Rovira-Más F (2016) Preliminary approach for real-time mapping of vineyards from an autonomous ground robot. ASABE Paper n° 162457331. ASABE, St. Joseph, USA

Saiz-Rubio V, Rovira-Más F, Millot C (2017) Performance improvement of a vineyard robot through its mechanical design. ASABE Paper n° 1701120. ASABE, St. Joseph, USA

Stoll M, Jones HG (2007) Thermal imaging as a viable tool for monitoring plant stress. Journal International des Sciences de la Vigne et du Vin 41(2):77–84

Subramanian V, Burks TF (2007) Autonomous vehicle turning in the headlands of citrus groves. ASABE Paper n° 071015. ASABE, St. Joseph, USA

USDA (2012) Census of Agriculture: Specialty Crops, vol 2. 2012. https://www.agcensususda.gov/Publications/2012/Online_Resources/Specialty_Crops/SCROPS.pdf. Accessed May 2018

VineRobot project (2014) Grant agreement 610953. www.vinerobot.eu. Accessed 2 May 2019

Wine Institute. http://www.wineinstitute.org/resources/statistics. Accessed May 2016

8camera. www.8camera.com. Accessed May 2016

Chapter 6
Robotic Spraying for Precision Crop Protection

Roberto Oberti and Ze'ev Schmilovitch

Abstract Plant protection products play a strategic role in securing worldwide food production. Nevertheless, major societal concerns are raised about risks for the environment and humans, and are being addressed by policy actions for reducing the dependence of agriculture on pesticides. A primary contribution to this can come from precision crop protection approaches, with treatments tailored to the site and time-specific needs of protecting the crop from pest pressure and expected infestation spreading. Robotic systems can play a role in precision crop protection, both in accurate monitoring of plant conditions and in timely and selectively spraying the treatment targets. This chapter provides a comprehensive discussion of the state-of-art of robotic spraying, with a review of weed and disease sensing tasks, and of precision actuation of treatments. Despite the complexity of some of the problems ahead, the chapter shows how the building blocks of integrated robotic systems for precision crop protection are developing fast.

6.1 Precision Crop Protection Concept

Plant protection products (PPPs), commonly referred to as pesticides, include herbicides, fungicides, insecticides and other groups of chemicals that are active against harmful organisms (weeds, fungi, pest insects and so on) for the establishment and healthy growth of the crop. The PPPs can be produced by both industrial processes or derived from natural ingredients, and are very often formulated for liquid distribution in form of droplets from spraying nozzles.

R. Oberti (✉)
Department of Agricultural and Environmental Science – DiSAA, Università degli Studi di Milano, via Celoria, 2-20133 Milan, Italy
e-mail: roberto.oberti@unimi.it

Z. Schmilovitch
Institute of Agricultural Engineering, Agricultural Research Organization, The Volcani Center–P.O.B 15159, Rishon Lezion 7505101, Israel

© Springer Nature Switzerland AG 2021   117
A. Bechar (ed.), *Innovation in Agricultural Robotics for Precision Agriculture*,
Progress in Precision Agriculture,
https://doi.org/10.1007/978-3-030-77036-5_6

It is evident that PPPs play a crucial role in securing worldwide production of healthy and abundant food (Oerke 2006; Cooper and Dobson 2007). On the other hand, their use and misuse represent a major societal concern about the negative impact of agriculture on the environment, on the safety of food and eventually on human health. In addition, cases of increasing resistance to pesticides by weeds and pathogens can lead to further problems in crop protection (Hawkins et al. 2018). To address these concerns, the reduction of dependence on pesticides is one of the main objectives of many policy actions related to agricultural sustainability around the world.

In current farming practice, PPPs are typically applied uniformly to fields, with treatments repeated at almost regular time intervals, aiming to obtain continuous protection of crop plants from pests and diseases. Despite this uniform approach, it is well known that weed pressure varies considerably with time and space within fields. Several pests and diseases also exhibit an uneven spatio-temporal distribution, with characteristic sparse patchy structures evolving around discrete foci (i.e. localized initial infection spots), especially at the early stages of development (Waggoner and Aylorf 2000; Gerhards 2010; Nutter et al. 2010; Everhart et al. 2013). This type of pattern opens great opportunities in the development of the precision crop protection concept, i.e. the application of precision agriculture (PA) to protection operations, with spatial and temporal variation of the treatments. This accords with the fundamental principles of PA distributing only *where*, *when* and *how much* is needed by the crop.

Early stages of precision crop protection can be considered as the introduction of automated spot spraying of herbicide for weed control (Felton and McCloy 1992; Paice et al. 1995; Slaughter et al. 1999). This essentially relies on the rapid switching of solenoid nozzles to enable the spatially intermittent (spot) spraying of weeds patches based on either infestation maps prepared offline or on real-time optical sensing. The concept of variable spraying treatments was later extended to protection treatments directed also to crop plants. For fungicide applications, in particular, the objective of saving pesticide was pursued by varying the spray rate to adapt it to the changing crop canopy, both in arable crops (Miller et al. 2000; Dammer and Ehlert 2006; Van De Zande et al. 2009; Dammer and Adamek 2012) and in orchards and vineyards (Balsari and Tamagnone 1998; Moltó et al. 2001; Solanelles et al. 2006; Gil et al. 2007) where the canopy volume largely changes during the growing season, and gaps in the vegetation can often occur.

Advancing towards full implementation of the precision crop protection concept requires that technological, biological, agronomic, epidemiological, agrochemical and modelling components have to be integrated to obtain a system that eventually will be able to (West et al. 2003; Oberti et al. 2014):

i.   sense crop conditions in the field at very high resolution,
ii.   detect and classify the early signals of infection symptoms within large volumes of sensed data,
iii. foresee the most likely spatio-temporal pattern of infection spreading by epidemiological modelling;

iv.   treat in a timely and selective way the initial foci (or the weed patches) and surrounding buffer areas threatened by spreading of the infestation.

While allowing significant savings in applied pesticide, such an approach would ensure that the healthy fraction of crop area is as large as possible by preventing the establishment and the following epidemic expansion of the infestation because of treatments targeted on the primary source of infection.

The reduction in pesticide obtained with this approach will depend on various factors, some of them inherently non-controllable. This is the case, for example, of the rate of pest or disease-specific expansion (i.e. the speed of the epidemic's gradient) that directly affects how large the buffer area should be around the treated detected foci. Or, again, the frequency of rainfall that can delay an application to a point that precision spraying will not be possible since the whole field needs protection as a result of the rapid spread of the infection.

The amount of pesticide reduction that is potentially possible with precision spraying depends strongly on the technical feasibility and economic viability of the solutions that will be implemented.

In this framework, robotic systems can play a fundamental role on two aspects of precision crop protection. Autonomous robots can enable:

i.   the patrol of fields to inspect the plant canopy or specific organs with on-board sensors and allow the monitoring of crop health at a very high spatio-temporal resolution by navigating along scouting paths that may adapt dynamically to previous findings and epidemiological modelling;
ii.   the timely treatment in the early stages of pest, disease or weed development by selectively applying the PPPs in field areas containing targets, i.e. single or groups of plants to be treated, or even to spray specific parts of plants (group of leaves, fruits, ears and so on), enabling an improved distribution of the pesticide on the organs that need to be protected.

During the last decade there have been specific advances in the area of crop protection with the impressive development of agricultural robotics, and despite the complexity of some of the problems, the building blocks of integrated robotic systems for precision crop protection are developing fast. Moreover, the first commercial robotic platforms for precision spraying are now available, showing the ongoing transfer of research achievements to operations in the field.

6.2   Robotic Sensing for Precision Protection Treatments

Robotic systems offer the inherent ability for automatic monitoring of crop conditions with a spatio-temporal resolution that outperforms the remotely-sensed measurements taken from airborne vehicles or satellites. Depending on the characteristics and design of sensors with a sensing distance in the range of 1 m from the plants, proximal measurements with a sub-millimetre spatial resolution can be achieved with current devices. This level of detail in the data recorded potentially enables deviations from a healthy status to be detected at a very early stage and at the single leaf scale. Moreover, robotic platforms can potentially be equipped with manipulators (robotic arms) designed to enable parts occluded from view to be measured, or to collect physical samples at specific points of the field for subsequent offline analysis with diagnostic methods (Menendez-Aponte et al. 2016).

On the other hand, the temporal resolution of measurements is also crucial for understanding the dynamics and rate of the infestation, and for the prompt triggering of protection treatments. To this end, the temporal information retrieved from remotely-sensed data can suffer from a limited rate of revisiting offered by satellite orbits and from possible cloud cover during critical stages of crop growth, in addition to covered plots or vertical canopies that cannot be measured properly from above.

Conversely, a robot can repeat the field scouting at close time intervals, depending on the total area to be covered. For example, considering a ground platform with a sensing swath of 2 m and a moderate forward speed of 0.3 m s^{-1}, able to operate night and day for, say, 18 h per day, it could revisit every single plant in a 16-ha field every 4 days. The scouting path of the robot might also adapt dynamically to the detection of weed spots or of pest or disease foci, for example by increasing the frequency of revisiting in critical areas of the field where spreading of the epidemic is likely to occur according to epidemiological modelling.

Use of aerial platforms allows dozens (or even hundreds) of times faster monitoring of crop areas, and the increasing availability of unmanned aerial vehicles (UAVs) suitable to carry high resolution sensors at low altitude (from a few metres to dozens of metres above the ground) along programmed paths over fields, is very promising for applications to precise monitoring of crop health at the farm scale.

This speed advantage is generating a growing interest for the use of UAVs in crop monitoring, particularly for tasks relying on top views and on millimetre-scale data resolution, as it is for weed detection. On the other hand, UAVs may have certain limitations compared to ground platforms, such as resolution capabilities below what is needed for the detection of very early disease symptoms, or the uncontrollability of measurement conditions (in the case of optical sensing: illumination, shadows and so on), or of reduced capability of inspecting the vertical canopy of trees and bush crops (vineyards, orchards etc.) (Yue et al. 2012; Torres-Sánchez et al. 2013; Hernández-Clemente et al. 2019).

In spite of the platform used and specific autonomous navigation system adopted, the fundamental sensing task for precision crop protection is to detect signals of biotic threats to crop health early, i.e. emergence of weeds and symptoms of disease or of insect pest infestation. To this aim, the most common sensing techniques used, by far, are based on the measurement of optical proprieties of plants. They are non-destructive, therefore, these measurements can be repeated without interfering with crop growth. In addition, they are compatible with on-the-go sensing without requiring any direct contact between instrument and plant and this also avoids the risk of spreading the disease.

Optical sensing relies on measuring properties of the radiation emerging from plant surfaces after multiple interactions, i.e. reflection, transmission and absorption, with tissues and organs of the plant. This reflected or re-emitted radiation forms the plant's spectral signature that, for crop monitoring purposes, is particularly relevant in the visible (VIS, wavelength range $\cong$ 400–700 nm), near-infrared (NIR, wavelength range $\cong$ 700–1100 nm), shortwave infrared (SWIR, wavelength range $\cong$ 1100–2500 nm) and thermal infrared (TIR, wavelength range $\cong$ 5–15 μm) bands.

The spectral signature of vegetation in these bands depends on (Jacquemoud and Ustin 2001; Baret et al. 2007):

i.   content of absorbing components in the tissue, mainly photoactive pigments (chlorophylls, anthocyanins, carotenoids), water, and to less extent proteins and other carbon constituents;
ii. canopy structure and morphology, i.e. the spatial arrangement, orientation and density of leaves, their texture and total area;
iii. surface temperature of the vegetation (for thermal infrared emission).

These characteristics depend, in turn, both on plant species and, all of the above, on the health status of the plant determined by the growing conditions, competition by weeds and pest or disease physiological disorders (West et al. 2003; Meroni et al. 2010).

6.2.1 Weed Sensing for Targeted Treatments

Based on the evident differences between vegetation and bare soil in the spectral reflectance in the visible (notably in the green and red bands) and near infrared zones, the automatic detection of weeds before crop sowing or planting is an immediate task. For this, commercial optoelectronic sensing systems are already available for sprayers to detect and to spray green areas directly in the field.

Robotic capabilities become essential to tackle the much more complex task of detecting weeds within vegetation. In this case, the discrimination of weeds from crop plants, particularly for seedlings, requires high-resolution measurements for single plant scale data processing.

Multivariate reflectance spectroscopy with hyperspectral line-imaging systems that allows spectral and spatial information to be combined, has been used in field studies for plant species recognition from a ground platform set-up which measures plants from a top view at a height of 0.5–1.5 m. The sensing systems have been operated both in sunlight and under controlled illumination, and after training a classifier algorithm. Accuracies in discriminating crop from weed plants in validation were typically 80–90% (Vrindts et al. 2002; Okamoto et al. 2007; Slaughter et al. 2008; Zhang et al. 2012; Herrmann et al. 2013).

Leaf shape, plant morphology and spatial positioning (e.g. random distribution in contrast to regular patterns or row crops) are the properties used by humans to identify weeds in the field. These spatial properties can be extracted and quantified by image processing, which to date appears the most promising technological approach to weed sensing.

Colour images in the red, green and blue (RGB) channels of the visible spectrum or multispectral images (typically including VIS and NIR bands) for weed sensing are often acquired under varying natural illumination or, less frequently, using enclosures with artificial lighting to provide uniform and constant illumination conditions on the field of view of the cameras. Images acquired under natural -i.e. varying- illumination, are often normalized by channel histogram equalization.

Accurate segmentation of vegetation pixels from the background, soil and non-living plant residues is usually obtained by applying the differences in spectral properties at the pixel level. For this, various spectral indices, i.e. the algebraic combinations of pixels' grey levels in two or more spectral channels, such as the normalized difference vegetation index (NDVI) or and the excess green index (ExG), have been used (Woebbecke et al. 1995; Guerrero et al. 2012; Lottes et al. 2016; Li et al. 2016).

At an early stage of growth or with low-density vegetation, the segmented foreground area often displays individual plant leaves, which can undergo morphological analysis. Simple geometric parameters, such as perimeter, area, equivalent diameter, minor and major axis and so on, can be easily computed for each connected region to characterize and possibly classify the shape of leaves (Berge et al. 2008; Kaspersen et al. 2010). Nevertheless, shape features that are less sensitive to scale, such as leaf orientation, angle of inclination, partial overlapping, are often applied for robust assessments of leaf shapes from field images. Fourier transform or elliptic descriptors of boundary contours (Neto et al. 2006), Hu's moment invariants, beam angle statistics are among the most used shape descriptors for the classification of plant morphology (Cope et al. 2012).

'Active contouring' is an alternative group of methods, where several contour templates of candidate plant species are geometrically rotated, scaled and deformed iteratively to match with the vegetation foreground boundaries extracted from the image (Sogaard 2005; Persson and Astrand 2008; Swain et al. 2011). Depending on the cover density and related overlaps among leaves and plants, active shape methods performed remarkably well when applied to typical field images of arable crops, often with an accuracy of above 90% in weed–crop discrimination.

When the crop has been planted precisely according to a fixed grid (as for most horticultural crops, maize (*Zea mays* L.), soya bean (*Glycine max* (L.) Merr.) etc.), this regularity in the spatial pattern can reinforce the detection of weeds in the inter- or intra-row regions, based on shape or size differences between crop and weeds plants (Tillett et al. 2001; Onyango and Marchant 2003; Hague et al. 2006; Jones et al. 2009; Liu et al. 2014).

Three-dimensional sensing by light detection and ranging (LiDAR) sensors or by time-of-flight (TOF) cameras, often coupled to a colour camera to generate RGB-depth images, can be used to discriminate crop and weeds plants based on their height and biomass volume (Piron et al. 2011; Andújar et al. 2016), or to reduce the shape–confusion effect due to overlapping leaves in regular images (Fig. 6.1). In 2-D top view imaging, overlaps are very common when plants develop in size and the leaves tend to aggregate in regions, which can be difficult to classify or analyse correctly.

Fig. 6.1 Depth image of carrot plants and weeds: brighter pixels (the taller carrot leaves) are closer to the top-viewing camera than the darker pixels (lower weeds foliage). The image shows the great potential of 3D approaches in crop–weed discrimination (*source* Piron et al. 2011)

To overcome the limitations in crop–weed discrimination for large degrees of infestation, Raja et al. (2019) introduced a novel approach named 'crop signalling' that uses different non-toxic fluorescent markers to signal crop plants permanently when growing in the field (Fig. 6.2). The unique fluorescence emission characteristic under UV or blue light illumination enabled reliable discrimination of lettuce (*Lactuca sativa* L.) and tomato (*Solanum lycopersicum* L.) plants from weeds

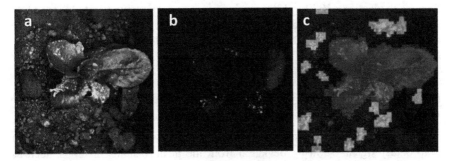

Fig. 6.2 Crop–weed discrimination based on non-toxic fluorescent markers for crop signalling when growing in the field: **a** marked lettuce plant within a weed patch, **b** imaged under UV excitation illumination, and **c** the precision weed and crop mapping created by the robot, where red colour indicates lettuce foliage and green colour indicates weeds (*source* Raja et al. 2019)

outdoors with large infestation densities, reaching an accuracy of detection of 99.7% for specific markers with no false positive errors.

Several classifiers have been used to combine and fuse image-extracted features to discriminate weeds within crop area. Readers interested in an exhaustive overview can see a recent review by Wang et al. (2019a) where several methods are listed and reviewed.

We mention here the deep learning approach of convolutional neural network (CNN) because they are being applied in almost any image-recognition related problem with remarkable results. The CNN can have different architectures, comprising convolutional (features extraction), pooling (dimensions reduction and fusion) and fully connected (classifiers) layers, with a final output layer of classified regions of input images (Kamilaris and Prenafeta-Boldú 2018). Adequate training of a CNN requires large sets of reference data, and at the end of the iterative process a network model is obtained for application. However, interpretability of the training results and of the influence of the network architecture used is poor (black box).

Applications of CNN to crop–weed classification are increasing rapidly. Dyrmann et al. (2016) trained and tested a CNN model derived from GoogLeNet on more than 10 000 images containing 22 weed and crop species at early growth stages, and acquired under different conditions of lighting, resolution and soil type. They obtained a classification accuracy of more than 85% for the 22 species considered. Potena et al. (2017) used a two-step CNN procedure for weed identification in images: a first simpler CNN was used for vegetation segmentation, and the extracted pixels were then classified as crop or weed by a second deeper CNN resulting in a classification precision close to 99%.

Olsen et al. (2019) collected and published DeepWeeds, a dataset of more than 17 000 labelled images, containing of eight weed species that are important for northern Australia. The dataset was used to train and validate two popular CNN architectures, obtaining for both an average classification accuracy above 95%. When operated on a dedicated high performance GPU card, the ResNet-50 model required on average 53.4 ms to classify a 1920 × 1200 pixels image, corresponding to a field area of 45 × 30 cm. These outstanding results are quite promising for implementation in the near future of robotic detection of weeds under real field conditions.

6.2.2 Disease Sensing for Targeted Treatments

Disease sensing under field conditions mostly relies on optical techniques. The aim is to detect signals showing deviation from the healthy status of crops, such as dysfunction or destruction of the photochemical pigments and modifications in plant tissue composition and structure, with corresponding appearance of chlorotic or necrotic lesions on leaves. These symptoms, with the possible addition of pathogen spores or propagules on the leaf surface, lead to increasing the tissue reflectance in the VIS range, especially in the red band where chlorophyll has a peak of absorption, and to shift the red-edge to shorter wavelengths and changing its slope. In addition, tissue senescence and reduced growth decreases the canopy NIR reflectance, changes in leaf water content modifies the spectrum in the SWIR range, and leaf temperature changes induced by modifications in the transpiration rate can be detected in the TIR band. (West et al. 2003; Hernández-Clemente et al. 2019).

Moreover, the emission of fluorescence in the far-red band by chlorophyll can be used to probe imbalances in the photochemical efficiency, i.e. in plant health status, so that increases in fluorescence intensity can indicate early stages of disease or other emerging stresses (Scholes and Rolfe 2009; Wright et al. 1995). These changes in spectral reflectance or in fluorescence emission do not provide unambiguous information of the presence of a specific stress, but rather a signal of non-normal conditions in the crop plants, as disease or other symptoms (West et al. 2003). Optical sensing offers great potential for the early detection of crop pests and diseases (i.e. emerging deviations from healthy status). Diagnostic capabilities (i.e. identification of symptoms of specific diseases), however, still require much development for robust applications even though many past investigations moved from the postulated idea of associating a specific stressor to unique spectral signatures (Nutter et al. 2010).

High-resolution spectral measurements have been successfully used for the detection, and in some specific cases the discrimination, of different diseases symptoms in grapevine (*Vitis vinifera* L.) (Naidu et al. 2009) wheat (*Triticum aestivum* L.) (Bauriegel et al. 2011), sugar beet (*Beta vulgaris* L.) (Mahlein et al. 2010), citrus (Pourreza et al. 2016), tomato (Wang et al. 2019b), and other crops.

Hyperspectral imaging provides information in both spatial and spectral dimensions of the imaged area, and potential use for disease detection was demonstrated under both laboratory and field conditions (Bravo et al. 2003; Larsolle and Muhammed 2007; Delalieux et al. 2007; Rumpf et al. 2010). More recent work on hyperspectral disease sensing has focused on advanced data mining techniques of hypercube data for automatic diagnostic capabilities (Wahabzada et al. 2015; Pantazi et al. 2017), including deep learning approaches (Polder et al. 2019; Wang et al. 2019b).

High-resolution imaging techniques have been applied extensively in this area, including the investigation of disease symptoms at very early stages. This is the case of fluorescence imaging under controlled conditions (Chaerle et al. 2004; Scholes and Rolfe 2009; Bellow et al. 2013; Buschmann et al. 2013) or in field applications (Bodria et al. 2002, Raesch et al. 2014, Šebela et al. 2014) with the use of a complex measurement setup (Pérez-Bueno et al. 2019), or thermography imaging for the detection of disease-induced modifications in tissue temperature because of the impairment of plant transpiration and water status (Oerke and Steiner 2010; Kaur et al. 2019).

The RGB imaging or multi-spectral imaging (which includes additional channels in NIR or other spectral bands), are probably the most widely studied techniques for disease sensing applications. Countless devices from mobile phones, pocket cameras, miniaturized multispectral cameras for UAVs, including high-resolution digital cameras are being used for image acquisition aimed at crop health monitoring. Interested readers can refer to recent extensive reviews on this topic (Sankaran et al. 2010; Martinelli et al. 2015; Mutka and Bart 2015; Barbedo 2016; Mahlein 2016).

Fig. 6.3 Multi-sensor detection of grapevine diseases from a mobile platform in vineyard. **a** multiple cameras measuring a portion of the canopy from within a tunnel enclosure, to obtain diffuse illumination and background regularization. **b** the corresponding R-G-NIR multispectral image; the yellow frame indicates the narrow vertical stripe of the canopy measured by an hyperspectral camera. **c** the corresponding hyperspectral image with the horizontal size representing the spectral axis (in this case, 425–900 nm), while the vertical size corresponds to the different points along the line of view of the camera. In the upper part of hyperspectral image can be seen the spectral intensity from one of the reflectance standard targets, enabling the normalisation of data acquired under different illumination conditions (source: photo of the author)

In general, particular attention must be given to the image acquisition setup, especially uniformity of the illumination that is crucial for the accuracy of automated image analysis. With this aim, even in the field, image acquisition is often carried out within enclosures or under light-diffusing shelters to obtain controlled illumination (Fig. 6.3). For certain diseases the view geometry of the camera has also proved to improve markedly the final detection of symptoms (Oberti et al. 2014).

Image processing and analysis aim to segment and classify the pattern of global or local features (pixels intensity, texture, shape and so on) associated with disease symptoms. This may be achieved with normalization or equalization, or by mathematical combination of different channels, i.e. by computing spectral indices, by applying simple or adaptive threshold, by computing intensity gradient, applying spatial filtering or domain transform etc. The features of regions extracted as candidate disease symptoms can then be analysed by several classification techniques, from the more classical linear discrimination analysis (LDA), principal components analysis (PCA), k-means or fuzzy C-means clustering, and more recently support vector machines (SVM), spectral angle mappers (SMA) and one of several variants of neural networks.

Recently, a burst of CNN applications to imaging-based detection and diagnostics of crop diseases has being published (Boulent et al. 2019). A lot of this deep-learning research used a public dataset (www.PlantVillage.org), now containing almost 90 000 colour images of leaves of 25 species of plants, healthy and infected, classified into 58 different classes (species × disease) acquired under very different conditions (www.

PlantVillage.org). In other cases, specialized and customized datasets are used to focus on a specific crop or disease (DeChant et al. 2017; Wiesner-Hanks et al. 2018). Disease images recorded under field conditions are typically acquired manually, from a ground vehicle or from a UAV, and so represent well the possible output from a robot platform scouting the crop health.

The positive results achieved by CNNs with reported disease detection accuracy often above 95% (DeChant et al. 2017; Brahimi et al. 2017; Lu et al. 2017) show the undoubted potential of deep learning techniques for robotic sensing and diagnostics of pests and diseases. Nevertheless, CNN models in crop health monitoring are still at initial, yet promising, stages of development and future experiments about training procedures on natural field data, transfer of learning and network architectures are needed to understand some unclear aspects better. For example, how network architecture influences the diagnostic capabilities, and why greater complexity and depth of the algorithm layers do not necessarily lead to greater accuracy (Fuentes et al. 2018; Brahimi et al. 2017), or why performance of the same network for similar tasks can vary significantly from one study to another (Fuentes et al. 2018; Too et al. 2019)

In quite a few notable cases, imaging systems were integrated on advanced autonomous platforms, exploring the feasibility of robotic disease sensing on-the-go under field conditions. For example, Polder et al. (2014) developed a fully enclosed field platform (Fig. 6.4) with multispectral and RGB cameras for mimicking humans

Fig. 6.4 Sensing platform for detecting tulip breaking virus (TBV) disease. The right inset shows the interior of the fully enclosed imaging area for obtaining controlled illumination (*source* Polder et al. 2014)

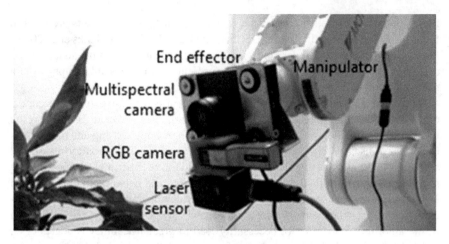

Fig. 6.5 Multiple sensors integrated in an end-effector and operated by a robotic manipulator to automatically inspect pepper plants in search of powdery mildew and of tomato spotted wilt virus diseases symptoms (*source* Schor et al. 2017)

in detecting tulip plants infected with tulip breaking virus (TBV). It was able to outperform human experts in the early detection of TBV symptoms with 90% accuracy and a 10% false positive rate. Schor et al. (2017) implemented and fully demonstrated a disease-monitoring robot for greenhouse pepper (*Capsicum frutescens* L.) plants, based on RGB and multispectral cameras and a laser beam distance sensor mounted on a robotic manipulator (Fig. 6.5). The system could detect powdery mildew and symptoms of tomato spotted wilt virus (TSWV) with an overall accuracy above 90% at a relatively fast inspection time of 26.7 s per plant on average.

Disease detection by moderately high-resolution imaging (at a scale of cm per pixel) from low-altitude flying aerial autonomous platforms is also being investigated. For this, lightweight unmanned aerial vehicles (UAVs) equipped with multispectral cameras were used for monitoring virus borne epidemics in grapevines, such as grapevine leaf stripe disease-GLSD (di Gennaro et al. 2016) and Flavescence dorée disease (Albetis et al. 2017), or fungal diseases in wheat (Su et al. 2019) and cucurbits (Kalischuk et al. 2019).

Bohnenkamp et al. (2019) conducted a comparative study on yellow rust detection in wheat by hyperspectral cameras mounted on ground and on UAV platforms. Because of the higher spatial resolution in the images (0.4 mm pixel^{-1} vs. 8 mm pixel^{-1}), the ground platform enabled much greater detection accuracy by applying the same processing and identification approach, leading the authors to conclude that technical advances in aerial stabilization and adopted optics are required in UAV platforms to achieve the sensitivity of ground platforms.

Among the non-optical techniques that are receiving research interest, volatile organic compounds (VOCs) profiling seems to be particularly promising for disease sensing applications (Cui et al. 2018; Li et al. 2019). The VOCs are molecules of metabolites released at very small gaseous concentrations by the tissue of a plant, with a composition, or profile, that can indicate the crop health status (Bos et al. 2013). Profiling of VOCs is routinely performed in the laboratory with bulky analytical instruments that are unsuitable for direct use in the field. This usually requires sampling of VOCs by small adsorbing probes that have to be exposed for enough time to the air surrounding the plant.

Profiling of VOCs from robots has not been explored yet, nevertheless there is a growing interest in implementing the portability of this approach into the field. There is a strong potential for its application on crop health monitoring robotic platforms (Duque Rodriguez et al. 2012). For example Fung et al. (2019), by modifying a commercial hand-vacuum base, developed a low-cost, mobile VOC sampling device equipped with solid phase micro-extraction (SPME) fibre probes. They demonstrated its use in measuring variation in representative plant in a citrus orchard, with a collection time of 5 min.

With their portability, electronic noses (e-nose) are receiving special attention for in-field VOC profiling. The e-nose device is an array of different gas sensors, mostly relying on electrical conductivity reversible changes in functional substrates with exposure to particular gases. Their data have been used successfully for monitoring pests and diseases in cucumber (*Cucumis sativus* L.), pepper, tomato, potatoes (*Solanum tuberosum* L.) and rice (*Oryza sativa* L.) plants in greenhouse or field controlled experiments (Laothawornkitkul et al. 2008; Jansen et al. 2009; Zhou and Wang 2011; Biondi et al. 2014; Cheng et al. 2017).

Similarly, precise detection of the presence of inoculum (as fungal spores) in air can be used as a warning of disease or target sensing (West et al. 2010). Current techniques are based on networks of stationary air samplers, able to trap airborne spores that are subsequently analysed with DNA-based diagnostic methods for pathogens pressure monitoring (Choudhury et al. 2017). Nevertheless, ground or aerial robotic platforms could be used for precise pest and pathogen sampling at different locations and in time to identify potential infection sources and population structure (West and Kimber 2015).

6.3   Precision Spraying Robotic Platforms

Current developments in robotic actuation of crop protection involve two approaches, with different requirements in terms of dynamics and dexterity capabilities of the components involved. One is robotic spraying, which aims to deliver selectively and deposit precisely an appropriate amount of liquid pesticide onto the detected targets (weeds, diseases, pest infestations). A second area is robotic weeding, which specifically aims at controlling or destroying weeds by physical means, such as mechanical, thermal or electrical treatments.

This section focuses on robotic spraying, and it will not cover mechanical weeding although it is an important topic for agricultural robotics (see Chaps. 3 and 7)).

6.3.1 System Architecture

Regardless of the platform on which they are implemented and different details of the specific configuration, spraying robots share a common system architecture that includes two main components.

One is the **navigation system** that controls and operates the task-oriented motion of the platform and, taking into account the environmental constraints, ensures that all the target positions planned in the spraying mission are reached. For this, it combines accurate positioning by GNSS, IMU and platform odometry, with range perception for local motion corrections and for reactive obstacle avoidance.

The second main component is the **spraying system** that controls and operates the pesticide delivery. When the robot, by moving along the planned route, reaches the treatment target sites specified by the operation mission plan (i.e. the prescription map), the spraying control triggers on the sprayer circuit. This comprises a main tank that can possibly be organized in a few multiple tanks with different pesticides, a volumetric pump, control valve(s) and spraying nozzles. Spraying control does not just enable setting of the on/off spray delivery, but also meters the distribution rate according to site-specific needs (e.g. canopy density, disease pressure and so on), or even to switch among applied pesticides (e.g. different fungicides or herbicides) during the treatment.

Spraying robots are often equipped with mechanical actuators or manipulators which can modify the nozzles' pose and distance to target to adapt the distribution pattern to the plant canopy shape or density and obtain improved pesticide deposition on targets.

The spraying system can also include a motion control component for the actuators adjusting the position or orientation of sprayer effectors, and sensing modules for the local perception of sprayed area and detection of treatment targets, e.g. weeds, diseased areas, group of leaves, fruits and other organs etc. (see previous Sect. 6.2).

Finally, wireless communication with a base control station enables remote interaction with the robot in the field, either by simple data retrieval or by sending mission updates with changes to the planned route or to spraying tasks that may include possible new findings by robot sensing in real-time (Fig. 6.6).

This data link can also serve inter-robot communication in fleet or swarm organization, or take manual remote control (teleoperation) of spraying. This can also be used to establish more advanced human–robot collaborative (HRC) approaches, for example with manual annotation of complex targets by remote experts via the internet, or by training the robot in classifying unpredicted situations.

6.3.2 Ground Robot Examples

Examples of field robotic platforms for crop spraying are quite limited at present in terms of operational and decisional autonomy. But a deeper and meaningful insight

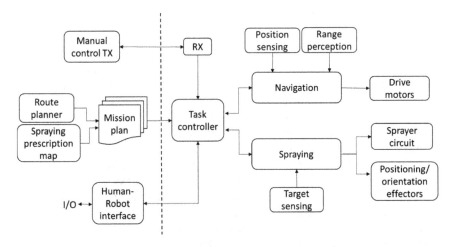

Fig. 6.6 General system architecture of a spraying robot. On the left of the dashed line: input tasks are provided from a ground station that can also enable it to tele-operate the robot manually or to establish an advanced cooperation with a remote expert through an internet based human–robot interface. On the right of the dashed line: the main components of the field robot

into this topic can be gained by considering studies on systems enabling advanced automation of spraying tasks, even if the equipment was coupled to a tractor or was driven manually. In this framework, studies on site-specific control for sprayers based on prescription maps in real-time (or offline) sensing of the targets, can be noted as useful examples of precursor research towards robotic spraying.

Such systems were developed for sprayers on **arable or row crops** to control the spray actuation of the whole boom or of individual sections, providing an operational spatial resolution the size of the spray boom, or larger than several metres. Compared to conventional uniform spraying, even with such limited resolution, site-specific spraying enabled potential savings in herbicide reported to vary from 5% to almost 90%, depending on the spatial and temporal distribution of weeds in the treated plots (Gerhards et al. 2002; Nordmeyer 2006; Christensen et al. 2009; Riar et al. 2011). Mink et al. (2018) reported a site-specific herbicide spraying system with a resolution of 9 m × 9 m gird cells based on application maps in maize and sugar beet derived from UAV image analysis. A similar site-specific approach has also been adopted for fungicide and insecticide treatments on arable crops. By applying sensor-controlled variable spray rate adapted to the changing crop canopy volume and density, pesticide reduction was reported to be 5 to 30% compared to conventional uniform spraying, while keeping the same average biological efficacy (Miller et al. 2000; Dammer and Ehlert 2006; Van De Zande et al. 2009; Dammer and Adamek 2012).

These precursor systems had recently evolved into robotic sprayers. Gonzalez-de-Soto et al. (2016) described the development and assessment of an autonomous robotized herbicide patch spraying system assembled on a commercial agricultural vehicle chassis, with a 6-m wide boom, direct-injection lines (separate tanks and circuits for different concentrated pesticides to be diluted in water just before spraying), and individual nozzle control by high-speed valves. The spraying robot was configured to execute herbicide treatments based on both a prescription map received from an external mission control or on real-time detection of weed targets in row crops by on-board cameras. In laboratory and field tests the robot showed a potential treatment accuracy of about 99.5% of the detected weeds, with a spatial resolution of 0.5 m by 0.5 m. This robot was developed in the RHEA project (see Chap. 7), where the concept of a cooperative fleet of multiple robots for crop protection (Fig. 6.7) was also explored (Gonzalez-de-Santos et al. 2017). Within this architecture, the autonomous sprayer represented the actuation node of a complex system with tasks coordinated by a mission manager base-station. The multi-robot system included UAVs as remote monitoring platforms with field inspection missions for weed detection. The remote data were processed in quasi real-time by the base station and used to plan (or modify) the mission to deploy to autonomous ground platforms, including the above boom sprayer robot (Gonzalez-de-Soto et al. 2016) and a highly automated olive (*Olea europaea* L.) orchard sprayer (Sarri et al. 2014), with continuous supervision to make real-time decisions if unexpected events occurred or to address safety issues.

Robotic technologies disclosed the potential implementation of ultra-precision protection treatments, with a sufficiently fine spraying resolution to treat individual seedlings or leaves selectively. In this regard, Lee et al. (1999) designed one of the first robotic micro-spraying systems with machine vision based technology for real time recognition of weeds in tomato crop rows and the capability of spraying resolution cell size of 1.25 cm by 0.63 cm. The micro-spraying system was constructed as a linear array of tubes each covering a width of 1.25 cm, individually controlled by fast solenoid valves that were tested at a forward speed of 0.7 km h^{-1} in an intra-row weed control experiment on tomatoes. Similarly, Nieuwenhuizen et al. (2010) reported a micro-spraying system consisting of needle nozzles each covering a working width

Fig. 6.7 The cooperative fleet of robots RHEA. The UAV platforms (on the right) have the task of remote monitoring of fields for weed detection; the processed data are sent to autonomous ground platforms (on the left) for physical weeding or selective spraying of herbicides (*source* Gonzalez-de-Santos et al. 2017)

of 4 cm, and used for controlling volunteer potatoes identified in sugar beet rows by machine vision. The system was coupled to a tractor operating at 3 km h^{-1}, and could control the growth of 77% of the weeds, while only 1% of the crop plants were damaged as a side-effect.

Sogaard and Lund (2007) integrated a similar micro-spraying apparatus on an autonomous robot vehicle able to navigate in a field with an accuracy within 5 cm. The micro-boom covered a 10-cm treatment swath by a linear array of 20 evenly spaced tubes, enabling a targeting precision of 5 mm by 5 mm with individual shots of liquid polymer herbicide formulation, which aimed to reduce the splash problem at the hit of the weed leaf. Jeon and Tian (2009) developed a field robot for direct herbicide application based on an ActivMedia Pioneer3 platform equipped with a 50-cm length, 5 degrees of freedom (DOF) arm, with a specific end effector based on two micro-pumps and a small circular saw. The end effector was designed to apply the chemical directly to the weed's cut surface by a sponge applicator. Small amounts of the chemical are wiped on to the cut stem exposing it to the vascular tissue. Field tests on two species of weeds resulted in 91% weeding effect with 78% reduction in pesticide reduction compared to a broadcast application.

Utstumo et al. (2018) developed a complete robotic system named Asterix, designed for micro-applications based on drop-on-demand of herbicide in row crops (Fig. 6.8). The three-wheeled robot was equipped with a linear array of needle nozzles covering a swath width of 17 cm with individual droplets of herbicide, spaced 6 mm apart. While navigating along a crop row, individual nozzles are activated by a machine vision based controller. In a field trial on carrots, the robot effectively controlled all the weeds with a reduction of herbicide use by more than 90% compared to label recommendations.

The operational achievements and savings in pesticide reported above have currently stimulated a range of implementations of modules for protection treatments on multipurpose field robots for row crops, or more rarely arable crops, such as the Robocrop Spot Sprayer by Garford (www.garford.com), BoniRob by Bosch (www.deepfield-robotics.com), Thorvald by Sagarobotics (www.sagarobotics.com), Dino by Naïo Technologies (www.naio-technologies.com), AVO by Ecorobotix (www.ecorobotix.com).

It is specialty crops for which precision spraying has a greater potential of pesticide savings, specifically on **bush and tree crops** where the total amount of pesticide used (application volume and frequency of the treatments) is typically much larger than for arable crops. Furthermore, bush and tree crops exhibit large differences in canopy volume and density during the season. Gaps in the vegetation or variation in canopy structure often occur within a field or between fields.

To address this heterogeneity of treatment targets, spraying control based on sensed canopy characteristics was developed not just for the on/off switching of individual nozzles but also to control the pattern of the spray in proportion to foliage density and according to canopy geometry (i.e. the plant's shape). Solanelles et al. (2006) and Chen et al. (2012) proposed an air-assisted sprayer for tree crops, fitted with a LiDAR sensor for canopy characterization and high-frequency PWM solenoid-operated nozzles. This enabled continuous variation of the flow rate delivered by each

Fig. 6.8 The weeding robot Asterix is an example of ultra-precision spraying robot. The individually controlled micro-spraying nozzles can apply herbicide selectively on sub-centimetre weed seedling targets (*source* Utstumo et al. 2018)

nozzle to adapt it to the foliage currently sprayed. This resulted in estimated pesticide savings of 25 to 45% compared to conventional treatments. Gil et al. (2013) and Román et al. (2020) developed similar systems for vineyard applications able to vary the distribution rate continuously at three different heights of the canopy with average savings above 20%.

Balsari et al. (2008) and Gil et al. (2015), in addition to the delivery rate of liquid spray, addressed the problem of controlling the air-assist flow rate to improve targeting and deposition of pesticide, by introducing adjustable air ports, individually controlled in three separate bands according to canopy volume. Vieri et al. (2013) went further by developing an orchard–vineyard sprayer able to control the inlet air flow rate and the delivering angle of four independent air ports to optimize the distribution pattern to canopy volume and shape sensed in real time by ultrasonic transducers.

Osterman et al. (2013) addressed the same objective by integrating a hydraulically driven robotic arm in an orchard sprayer, which enabled 8 DOF when configuring the spraying and air-assist pattern on the side of a tree row. The canopy shape was sensed by LiDAR and processed in real time to adjust the pose of the air-assist and spray delivery devices.

Berenstein et al. (2010) experimented with an autonomous robotic sprayer for vineyards, designed to treat selective grape clusters. The robot control, based on a machine vision detection algorithm, was able to detect the exact location of 90% of grape clusters and to target spraying towards them during vineyard tests. With this selective capability, and to the envisaged adoption of pan/tilt control of spray nozzles, the authors estimated a potential reduction of almost 50% in pesticide used for cluster protection.

As a further step in the concept of precision plant protection, the objective of selectively targeting pesticide on diseased areas in specialty crops has recently gained interest among researchers. Oberti et al. (2016) reported the first experiment of fully automatic selective spraying of diseased areas in speciality crops with a reconfigurable, multifunction agricultural robot developed within the project CROPS. They used a 6-DOF robotic manipulator equipped with a precision-spraying end-effector (Fig. 6.9), and an integrated multispectral imaging disease-sensing system, enabling the control of the actuation and pose of the robotic sprayer. The experiments were conducted on grapevines with different degrees of powdery mildew disease. The robot was able to detect and spray 85–100% of the diseased area within the canopy, with reduced pesticide use from 65 to 85% (depending on disease levels and spatial distribution of symptoms) when compared to conventional spraying. The precision-spraying end-effector (Malneršič et al. 2016) of the CROPS robot was designed as

Fig. 6.9 The multifunction agricultural robot CROPS equipped with a precision-spraying end-effector during an experiment of selective treatments on powdery mildew disease in grapevine. This is among the first examples of fully autonomous disease treatment, with selective spraying targeted on symptoms detected by an on-board sensing system (*source* Oberti et al. 2016)

a compact, air-assisted, high-resolution spot-sprayer unit able to treat target structures with typical size of 15–20 cm (i.e. one leaf or a group of small leaves) from a distance of 1–1.5 m. To improve cover of both leaf sides, the spraying effector delivered airflow speed pulses to produce local turbulence at the target distance and to induce leaf movement while spraying.

An interesting approach for enhancing the functional and selective capabilities of autonomous spraying robots is human–robot collaboration, where the robot operates mostly autonomously but for some tasks or in the case of unexpected events it interacts with a remote human operator. In this framework, Adamides et al. (2017) designed different interfaces for human–robot interaction devices with autonomous spraying robots and evaluated the perceived usability of a tele-operated agricultural sprayer in a series of field experiments in vineyard rows.

Berenstein and Edan (2017) reported an experiment on a human–robot collaborative autonomous sprayer for selective treatments on grape cluster targets. They developed a robotic platform equipped for autonomous and human-assisted field tasks such as navigation along the vineyard row and accurate spraying at the target area. The human and robot work collaboratively to detect targets in real-time field images acquired by the robotic platform. Depending on the collaboration level (manual target marking; robot suggests, human approves, robot marks, human supervises, fully autonomous robot), the system request, the human at a remote location to assist in detecting the spray targets from a computer console. In a simplified field experiment, the study showed the technical feasibility of human–robot collaboration in vineyard spraying with a potential reduction of sprayed material and evident improvement in detecting spraying targets both in terms of True and False Positive rates.

6.3.3   Aerial Platforms Examples

Lightweight unmanned aerial vehicles (UAVs) have been increasingly used for agricultural applications in the last decade. Commonly, remotely controlled semi-autonomous UAVs (often-named drones) are used in aerial monitoring of crop conditions, but mostly operating as passive devices for field data collection.

Interestingly, spraying is one of the few agricultural applications where UAVs are used to interact and operate actively in the field. Despite the limited payload capacity, aerial semiautonomous or robotic systems can potentially apply pesticide over fields, avoiding soil compaction or crop damage when treating tall plants (e.g. maize, sunflower (*Helianthus* L.) and so on), or on terraced plots in sloping areas where the limited trafficability allows only manual treatments by knapsack sprayers.

Because of safety regulations, UAV spraying platforms can carry a payload of 10–20 kg and have battery powered multiple rotors (multicopter), although some petrol engine helicopters exist (Yang et al. 2018). One or more nozzles are generally mounted below the rotors, delivering very small volumes (typically, 5–50 L ha^{-1}) of

Fig. 6.10 Lightweight unmanned aerial vehicles (UAVs) are used extensively in crop monitoring, but with their advanced automation they can potentially operate as aerial platforms for precision spraying. However, in most countries spraying by UAVs is strictly regulated or completely banne because of limited knowledge about pesticide drift from UAVs (in the photograph: a DJI Agras MG-1 spray drone. *Source* DJI.com)

spray carried to the ground by the airflow generated from the UAV rotors, generally operating at an altitude of 1–2 m and with a moderate spraying speed of 2–3 m s^{-1}.

Development of UAV spraying (Fig. 6.10) has been especially strong in East Asian countries leading to parallel commercialization of a myriad of specialized platforms, some are known such as the DJI-MG series (DJI Co, Shenzhen, China), the XAG-P series (XAG Co, Guangzhou, China) or the Yamaha-RMAX (Yamaha Motor Co, Iwata, Japan).

Research on UAV spraying has focused on aspects that include: the development or optimization of technological components to be integrated on commercial drone platforms, the implementation of autonomous capabilities of the platform, the testing of pesticide distribution and crop protection quality.

Among the first ones, Zhu et al. (2010) introduced a spraying controller based on a pulse width modulation (PWM) that was integrated with the guidance system of a UAV helicopter, enabling the pesticide volume to be delivered according to a pre-programmed treatments map. Xue et al. (2016) developed an automatic navigation control for a UAV spraying system operation, which integrated the flight plan route with a prescription map for spraying. In field tests the system reached sub-metre precision in following the designated spray routes, with spraying uniformities within 25% of the coefficient of variation under wind speed of 2 m s^{-1}, and using a swath width of 7 m.

Full autonomous capabilities of UAV spraying were explored by Dai et al. (2017), who developed a whole control system able to recognize targets and spray them without any human intervention. The on-board system, that was tested on artificial targets, includes navigation, vision-based target recognition and spraying controls all coordinated in real time by a task scheduler of the basic states, namely pre-searching of targets in the scene, target identification, spraying.

Autonomous UAV spraying requires accurate task scheduling (flight to targets, spray, tank refill, battery recharge and so on) and path optimization because the operation is constrained by limited energy and payload capacity. On-going research on this applies heuristic approaches or optimization methods to UAV flight planning, as mixed integer linear programming (MILP), genetic algorithms (GA) and particle swarm optimization (PSO) (Kim and Morrison 2014; Ramirez-Atencia et al. 2017). Faical et al. (2017) proposed a collaborative task control approach, by integrating UAV control with data on local weather conditions acquired by ground sensors to enable automatic real-time correction of route and sequence of UAV actions, based on changes in wind speed and direction to avoid unsprayed areas or reduce pesticide drift.

Limited knowledge on the amount of drift (unwanted deposition on to off-target areas, i.e. other crops, water, soil, animals, humans and so on) that can be expected by UAV spraying is an important concern, and most countries are currently applying strict regulations or an almost complete ban of UAV pesticide treatments.

To identify optimal solutions to mitigate drift potential and to assist regulatory decision-making by public agencies, research is on-going on modelling how the spraying pattern and potential drift depend on UAV operating characteristics (speed, rotors, geometry and so on) and on environmental conditions (air temperature and humidity, and especially wind speed). Therefore, 3-D simulations by different computational fluid-dynamics (CFD) approaches have been carried out to evaluate the airflow field generated by specific rotor geometry and how the spray droplets are transported by downstream flow (e.g. Zhang et al. 2016; Teske et al. 2018; Wen et al. 2019a). A major effect on drift potential in multirotor UAV is linked with flight speed (that should be limited to few m s^{-1}) and spraying altitude (that should be as close as possible to the target, in the range of 0.5 m).

Complementary experimental measurements of UAV spray deposit have been conducted in indoor facilities (Zhang et al. 2018; Wen et al. 2019b) where operational conditions (speed, altitude, wind and so on) can be controlled. A few studies have also been conducted in extensive field experiments to evaluate pesticide deposit quality on both arable and tree crops (Brown and Giles 2018; Sarri et al. 2019; Wen et al. 2019a), where deposition samples from plants are collected by different methods. Early studies on grapevines showed that with low flight height and under recommended wind speeds within 3 m s^{-1}, the off-target drift was limited and that the measured ground deposition due to drift was much less than typically observed in aerial spraying of orchards and vineyards, and also less than certain configurations of ground sprayers in vineyards. On the other hand, UAV spraying can result in poor deposit on the underside (abaxial) of leaves when flight height cannot be very close to the canopy, leading to poor protection effects compared to ground sprayers, requiring further advances in operative optimization.

Even if technical advances are stimulating new regulations, it must be noted that at present UAV spraying of plant protection products is treated as aerial spraying which operate under different laws: e.g. in Europe it is currently banned, with very limited exceptions where there are no viable alternatives to the use of aerial spraying; in the USA it is allowed when it complies with product label specifications, and under pre-filing of detailed flight plans; in China it seems to be treated as a regular crop protection technique.

6.4 Economics and Conclusions

This overview shows that robotic precision crop protection is a blossoming area of research and development specifically on two main issues: (i) detecting specific protection needs by monitoring the crop at a very high spatio-temporal resolution and (ii) timely and selective spraying of treatment targets at their earliest stages of development.

Table 6.1 summarizes a selection of research on automation and robotics for precision crop spraying according to three categories: (1) the technological implementation, (2) the agricultural objective and 3) the crop type.

On-going advances in reliable weed–crop discrimination based on imaging techniques, and the parallel availability of high-resolution, fast control spraying technologies, are now enabling complete autonomous systems for selective robotic treatment of weeds, with a strong potential for reducing herbicide use.

On the other hand, while canopy-optimized distribution of fungicide and insecticide is already implemented, research on selective spraying of pests and disease is still at an early stage of development and remains a huge potential for improvement. This is mostly due to the limitations still encountered in automated detection systems for disease symptoms, mostly based on a single measurement technique. This might benefit from data-fusion by multiple sensors, including non-optical devices, and by collaborative frameworks where human expertise is available on demand to robotic platforms.

In addition, the more complex relation between detected symptoms and potential epidemic expansion requires the integration of pest- or disease-specific epidemiological modelling into the robotic mission planner to create spraying maps that also include the asymptomatic buffer area to be treated.

The examples of ground and aerial robotic platforms developed suggest that, when a human operator is not needed anymore, the most likely scenarios of application will rely on fleets or swarms of small systems with modular architecture for multiple operations, giving advantages in term of improved mobility and reduced compaction, flexibility and reliance against breakdown.

A final but fundamental remark is on the economics of such systems, which still lacks comprehensive findings (Lowenberg-DeBoer et al. 2020). Some indications

Table 6.1 Summary of selected research on automation and robotics for precision crop spraying. For an exhaustive review the reader should refer to the main text of the chapter

Technological implementation	Main objective	Crop	References
Spot spraying by on-off switching nozzles	Weed patches control	Arable and row crops	Felton and McCloy (1992), Paice et al. (1995), Slaughter et al. (1999)
Spray rate proportional to crop canopy	Optimal volume of fungicide	Arable crops	Miller et al. (2000), Dammer and Ehlert (2006), Van De Zande et al. (2009), Dammer and Adamek (2012)
	Avoid spraying canopy Optimal volume of fungicide	Orchards and vineyards	Balsari and Tamagnone (1998), Moltó et al. (2001), Solanelles et al. (2006), Gil et al. (2007)
Automatic geometry of spraying and air carrier	Optimal distribution of pesticide adapted to canopy volume	Orchards and vineyards	Balsari et al. (2008), Vieri et al. (2013)
Reflectance spectroscopy, Hyperspectral imaging	Crop–weed discrimination Weed detection	Arable crops, row crops, horticultural crops	Vrindts et al. (2002), Okamoto et al. (2007), Slaughter et al. (2008), Zhang et al. (2012), Herrmann et al. (2013)
Leaf shape/geometry image analysis			Woebbecke et al. (1995), Sogaard (2005), Neto et al. (2006), Persson and Astrand (2008), Berge et al. (2008), Kaspersen et al. (2010), Cope et al. (2012), Guerrero et al. (2012), Lottes et al. (2016), Li et al. (2016)
Plants spatial pattern image analysis		Row crops, horticultural crops	Tillett et al. (2001), Onyango and Marchant (2003), Hague et al. (2006), Jones et al. (2009), Liu et al. (2014)
LiDAR, 3-D image analysis		Arable crops, horticultural crops	Piron et al. (2011), Andújar et al. (2016)
Crop signalling with fluorescent markers		Horticultural crops	Raja et al. (2019)

(continued)

Table 6.1 (continued)

Technological implementation	Main objective	Crop	References
Convolutional neural network (CNN) image analysis		Arable crops, row crops, horticultural crops	Kamilaris and Prenafeta-Boldú (2018), Dyrmann et al. (2016), Olsen et al. (2019)
Fluorescence sensing/imaging	Early detection of fungal disease	Wheat, sugar beet	Wright et al. (1995), Bodria et al. (2002), Chaerle et al. (2004), Scholes and Rolfe (2009), Bellow et al. (2013), Buschmann et al. (2013), Raesch et al. (2014), Šebela et al. (2014), Pérez-Bueno et al. (2019)
Reflectance spectroscopy, Hyperspectral imaging	Virus and fungal disease detection under field conditions	Wheat, sugar beet, tomato, pepper, grapevine, citrus	Bravo et al. (2003), Larsolle and Muhammed (2007), Delalieux et al. (2007), Naidu et al. (2009), Rumpf et al. (2010), Bauriegel et al. (2011), Mahlein et al. (2010), Wahabzada et al. (2015), Pantazi et al. (2017), Pourreza et al. (2016), DeChant et al. (2017), Brahimi et al. (2017), Lu et al. (2017), Polder et al. (2019), Wang et al. (2019b)
RGB/Multispectral imaging	Virus and fungal disease detection under field conditions	Wheat, grapevine, pepper, tulip, orchard	West et al. (2003), Oberti et al. (2014), Polder et al. (2014), Barbedo (2016), Schor et al. (2017), DeChant et al. (2017), Lu et al. (2017), Wiesner-Hanks et al. (2018)

(continued)

Table 6.1 (continued)

Technological implementation	Main objective	Crop	References
Olfactometry, volatile organic compounds (VOCs) profiling	Fungal diseases detection under field conditions	Cucumber, pepper, tomato, potatoes and rice plants	Laothawornkitkul et al. (2008), Jansen et al. (2009), Zhou and Wang (2011), Duque Rodriguez et al. (2012), Bos et al. (2013), Biondi et al. (2014), Cheng et al. (2017), Cui et al. (2018), Li et al. (2019), Fung et al. (2019)
Ground platform for automatic disease detection	Virus and fungal disease detection in field	Wheat, grapevine, pepper, tulip	Moshou et al. (2011), Oberti et al. (2014), Polder et al. (2014), Schor et al. (2017)
Aerial UAV platform for automatic disease detection		Wheat, grapevine, orchard; cucurbits	Yue et al. (2012), Torres-Sánchez et al. (2013), Di Gennaro et al. (2016), Albetis et al. (2017), Su et al. (2019), Kalischuk et al. (2019), Bohnenkamp et al. (2019)
Robotic micro-spraying	Ultra-precision spraying of weeds	Tomato, Potato, lettuce, carrot	Lee et al. (1999), Sogaard and Lund (2007), Jeon and Tian (2009), Nieuwenhuizen et al. (2010), Utstumo et al. (2018)
Robotic spraying	Weed spraying, canopy adapted spraying	Corn, grapevine, olive (*Malus domestica* Borkh., 1803), apple	Osterman et al. (2013), Sarri et al. (2014), Gonzalez-de-Soto et al. (2016), Gonzalez-de-Santos et al. (2017)
Robotic selective spraying of disease	Targeted spraying of fungal disease-	Grapevine	Oberti et al. (2016)
UAV precision spraying	Map-based aerial spraying	Rice, grapevine	Zhu et al. (2010), Xue et al. (2016), Dai et al. (2017)
Human–robot collaboration	Expert assistance to field robot in classifying spraying targets	Vineyard	Berenstein and Edan (2017), Adamides et al. (2017)

about the estimated profitability of robotic spraying can be derived from Pedersen et al. (2006) who considered robotic micro-spraying on horticultural crops for weed control. By assuming a purchase cost of €65 000 for the robot and a farming area of 80 ha, they estimated a total operating cost of €260 ha^{-1}yr^{-1}, resulting in a cost saving of 15% compared to conventional boom spraying and manual inter-row hoeing.

Tona et al. (2018) conducted an economic analysis on different technologies for spraying equipment on vineyards and apple orchards, considering the very intense protection protocols against fungal diseases adopted in Central-Southern Europe. Their analysis also included an example of an autonomous robotic platform able to detect and selectively spray the diseased areas. For a total area above 10 ha, they estimated that the purchase price that would make the robotic platform profitable over conventional sprayers was €55 000 and €67 000 for grapevine and apple crops, respectively. The authors concluded that such a low market price-threshold for profitability was too challenging for the current industrial cost, and that savings in pesticide (and labour) cannot be the only factor for possible adoption of intelligent robotic platforms for precision spraying against simpler technologies.

The above emphasizes the need for systemic research to address the potential economic impact of robotic precision spraying beyond the farm gate, including the potential environmental and social benefits, and the value of crop data continuously acquired by the robot (Lowenberg-DeBoer et al. 2020).

References

Adamides G, Katsanos C, Parmet Y, Christou G, Xenos M, Hadzilacos T, Edan Y (2017) HRI usability evaluation of interaction modes for a teleoperated agricultural robotic sprayer. Appl Ergon 62:237–246

Albetis J, Duthoit S, Guttler F, Jacquin A, Goulard M, Poilvé H, Féret JB, Dedieu G (2017) Detection of Flavescence dorée grapevine disease using Unmanned Aerial Vehicle (UAV) multispectral imagery. Rem Sens 9: art. no. 308

Andújar D, Dorado J, Fernández-Quintanilla C, Ribeiro A (2016) An approach to the use of depth cameras for weed volume estimation. Sensors 16:972

Balsari P, Doruchowski G, Marucco P, Tamagnone M, Van De Zande J, Wenneker M (2008) A system for adjusting the spray application to the target characteristics. Agric Eng Int CIGR J 10:1–11

Balsari P, Tamagnone M (1998) An ultrasonic airblast sprayer. In: Proc of Eur Ageng international conference of agricultural engineering. EurAgeng 1998, Oslo, Norway. pp 585–586

Barbedo JGA (2016) A review on the main challenges in automatic plant disease identification based on visible range images. Biosyst Eng 144:52–60

Baret F, Houlès V, Guèrif M (2007) Quantification of plant stress using remote sensing observations and crop models: the case of nitrogen management. J Exp Bot 58:869–880

Bauriegel E, Giebel A, Geyer M, Schmidt U, Herppich WB (2011) Early detection of Fusarium infection in wheat using hyper-spectral imaging. Comput Electron Agric 75:304–312

Bellow S, Latouche G, Brown SC, Poutaraud A, Cerovic ZG (2013) Optical detection of downy mildew in grapevine leaves: daily kinetics of autofluorescence upon infection. J Exp Botany 64:333–341

Berenstein R, Shahar OB, Shapiro A, Edan Y (2010) Grape clusters and foliage detection algorithms for autonomous selective vineyard sprayer. Intel Serv Robot 3:233–243

Berenstein R, Edan Y (2017) Human-robot collaborative site-specific sprayer. J Field Robot 34:1519–1530

Berge T, Aastveit A, Fykse H (2008) Evaluation of an algorithm for automatic detection of broad-leaved weeds in spring cereals. Precision Agric 9:391–405

Biondi E, Blasioli S, Galeone A, Spinelli F, Cellini A, Lucchese C, Braschi I (2014) Detection of potato brown rot and ring rot by electronic nose: From laboratory to real scale. Talanta 129:422–430

Bodria L, Fiala M, Oberti R, Naldi E (2002) Chlorophyll fluorescence sensing for early detection of crop's diseases symptoms. Proc of Am Soc Agric Eng-CIGR World Congr 2002, Chicago, USA

Bohnenkamp D, Behmann J, Mahlein AK (2019) In-field detection of Yellow Rust in Wheat on the Ground Canopy and UAV Scale. Rem Sens 11: art. no. 2495

Bos LDJ, Sterk PJ, Schultz MJ (2013) Volatile metabolites of pathogens: a systematic review. PLoS Pathog 9:

Boulent J, Foucher S, Théau J, St-Charles PL (2019) Convolutional neural networks for the automatic Identification of plant diseases. Front Plant Sci 10:941

Brahimi M, Boukhalfa K, Moussaoui A (2017) Deep learning for tomato diseases: classification and symptoms visualization. Appl Artif Intell 31:1–17

Bravo C, Moshou D, West J, McCartney A, Ramon H (2003) Early disease detection in wheat fields using spectral reflectance. Biosyst Eng 84:137–145

Brown CR, Giles DK (2018) Measurement of pesticide drift from unmanned aerial vehicle application to a vineyard. Trans ASABE 61:1539–1546

Buschmann C, Konanz S, Zhou M, Lenk S, Kocsányi L, Barócsi A (2013) Excitation kinetics of chlorophyll fluorescence during light-induced greening and establishment of photosynthetic activity of barley seedlings. Photosynthetica 51:221–230

Chaerle L, Hagenbeek D, De Bruyne E, Valcke R, Van Der Straeten D (2004) Thermal and chlorophyll-fluorescence imaging distinguish plant-pathogen interactions at an early stage. Plant Cell Physiol 45:887–896

Chen Y, Zhu H, Ozkan HE (2012) Development of a variable rate sprayer with laser scanning sensor to synchronize spray outputs to tree structures. Trans ASABE 55:773–781

Cheng SM, Wang J, Wang YW, Wei ZB (2017) Discrimination of different types damage of tomato seedling by electronic nose. ITM Web Conf 11:1–8

Choudhury RA, Koike ST, Fox AD, Anchieta A, Subbarao KV, Klosterman SJ, McRoberts N (2017) Spatiotemporal patterns in the airborne dispersal of spinach downy mildew. Phytopathology 107:50–58

Christensen S, Søgaard HT, Kudsk P, Nørrmark M, Lund I, Nadimi ES, Jørgensen R (2009) Site-specific weed control technologies. Weed Res 49:233–241

Cooper J, Dobson H (2007) The benefits of pesticides to mankind and the environment. Crop Protection 26:1337–1348

Cope JS, Corney D, Clark JY, Remagnino P, Wilkin P (2012) Plant species identification using digital morphometrics: a review. Expert Syst Appl 39:7562–7573

Cui S, Ling P, Zhu H, Keener HM (2018) Plant pest detection using an artificial nose system: A review. Sensors 18: art. no. 378

Dai B, He Y, Gu F, Yang L, Han J, Xu W (2017) A vision-based autonomous aerial spray system for precision agriculture. In: 2017 ieee international conference on robotics and biomimetics (ROBIO), pp 507–513

Dammer KH, Adamek R (2012) Sensor-based insecticide spraying to control cereal aphids and preserve lady beetles. Agron J 104:1694

Dammer KH, Ehlert D (2006) Variable-rate fungicide spraying in cereals using a plant cover sensor. Precision Agric 7:137–148

DeChant C, Wiesner-Hanks T, Chen S, Stewart EL, Yosinski J, Gore MA, Nelson RJ, Lipson H (2017) Automated identification of northern leaf blight-infected maize plants from field imagery using deep learning. Phytopathology 107:1426–1432

Delalieux S, van Aardt J, Keulemans W, Schrevens E, Coppin P (2007) Detection of biotic stress (Venturia inaequalis) in apple trees using hyperspectral data: Non-parametric statistical approaches and physiological implications. Europ Jour Agronomy 27:130–143

di Gennaro SF, Battiston E, di Marco S, Facini O, Matese A, Nocentini M, Palliotti A, Mugnai L (2016) Unmanned Aerial Vehicle (UAV)-based remote sensing to monitor grapevine leaf stripe disease within a vineyard affected by esca complex. Phytopath Mediterranea 55:262–275

Duque Rodríguez J, Gutiérrez López J, Méndez Fuentes V, Barreiro Elorza P, Gómez-Ullate D, Mejía-Monasterio C (2012) Search strategies and the automated control of plant diseases. In: Proceedings of 1st International conference on robotics and associated high-technologies and equipment for agriculture (RHEA). Pisa, Italy Sept 19–21, 2012, pp 163–170

Dyrmann M, Karstoft H, Midtiby HS (2016) Plant species classification using deep convolutional neural network. Biosys Eng 151:72–80

Everhart SE, Askew A, Seymour L, Scherm H (2013) Spatio-temporal patterns of pre-harvest brown rot epidemics within individual peach tree canopies. Eur J Plant Pathol 135:499–508

Faical BS, Freitas H, Gomes PH, Mano LY, Pessin G, de Carvalho AC, Krishnamachari B, Ueyama J (2017) An adaptive approach for uav-based pesticide spraying in dynamic environments. Comput Electron Agric 138:210–223

Felton W, McCloy K (1992) Spot spraying. Agric Eng (Nov.):9–12

Fuentes AF, Yoon S, Lee J, Park DS (2018) High-performance deep neural network-based tomato plant diseases and pests diagnosis system with refinement filter bank. Front Plant Sci 9, art. no. 1162

Fung AG, Yamaguchi MS, McCartney MM, Aksenov AA, Pasamontes A (2019) Davis CE (2019) SPME-based mobile field device for active sampling of volatiles. Microchem J 146:407–413

Gerhards R (2010) Spatial and temporal dynamics of weed populations. In: Oerke EC et al (eds) Precision crop protection – the challenge and use of heterogeneity. Springer, Dordrecht, pp 17–26

Gerhards R, Sökefeld M, Timmermann C, Kühbauch W, Williams MM (2002) Site-specific weed control in maize, sugar beet, winter wheat, and winter barley. Precision Agric 3:25–35

Gil E, Escolà A, Rosell J, Planas S, Val L (2007) Variable rate application of plant protection products in vineyard using ultrasonic sensors. Crop Protect 26:1287–1297

Gil E, Llop J, Gallart M, Valera M, Llorens J (2015) Design and evaluation of a manual device for air flow rate adjustment in spray application in vineyards. A: workshop on spray application techniques in fruit growing. In: Proceedings of the Suprofruit 2015—13th workshop on spray application in fruit growing. Linday, p 8–9

Gil E, Llorens J, Llop J, Fabregas X, Escola A, Rossel-Polo JR (2013) Variable rate sprayer. Part 2—vineyard protorype: design, implementation and validation. Comput Electron Agric 95:136–150

Gonzalez-de-Santos P, Ribeiro A, Fernandez-Quintanilla C, Lopez-Granados F, Brandstoetter M, Tomic S, Pedrazzi S, Peruzzi A, Pajares G, Kaplanis G, Perez-Ruiz M, Valero C, del Cerro J, Vieri M, Rabatel G, Debilde B (2017) Fleets of robots for environmentally-safe pest control in agriculture. Precision Agric 18:574–614

Gonzalez-de-Soto M, Emmi L, Perez-Ruiz M, Aguera J, Gonzalez-de-Santos P (2016) Autonomous systems for precise spraying – evaluation of a robotised patch sprayer. Biosyst Eng 146:165–182

Guerrero JM, Pajares G, Montalvo M, Romeo J, Guijarro M (2012) Support Vector Machines for crop/weeds identification in maize fields. Expert Syst Appl 39:11149–11155

Hague T, Tillett ND, Wheeler H (2006) Automated crop and weed monitoring in widely spaced cereals. Precision Agric 7:21–32

Hawkins NJ, Bass C, Dixon A, Neve P (2018) The evolutionary origins of pesticide resistance. Biol Rev Camb Philos Soc 94:135–155

Hernández-Clemente R, Hornero A, Mottus M et al (2019) Early diagnosis of vegetation health from high-resolution hyperspectral and thermal imagery: lessons learned from empirical relationships and radiative transfer modelling. Curr Forestry Rep 5:169–183

Herrmann I, Shapira U, Kinast S, Karnieli A, Bonfil DJ (2013) Ground-level hyperspectral imagery for detecting weeds in wheat fields. Prec Agric 14:637–659

Jacquemoud S, Ustin SL (2001) Leaf optical properties: a state of the art. In: Proc Int Symp Phys Meas Sign Rem Sens, pp 223–232

Jansen RMC, Hofstee JW, Wildt J, Verstappen FWA, Bouwmeester HJ, Posthumus MA, Van Henten EJ (2009) Health monitoring of plants by their emitted volatiles: Trichome damage and cell membrane damage are detectable at greenhouse scale. Annals Appl Biol 154:441–452

Jeon HY, Tian LF (2009) Direct application end effector for a precise weed control robot. Biosyst Eng 104:458–464

Jones G, Gée C, Truchetet F (2009) Assessment of an inter-row weed infestation rate on simulated agronomic images. Comput Electron Agric 67:43–50

Kalischuk M, Paret ML, Freeman JH, Raj D, Silva SD, Eubanks S, Wiggins DJ, Lollar M, Marois JJ, Charles Mellinger H, Das J (2019) An improved crop scouting technique incorporating unmanned aerial vehicle-assisted multispectral crop imaging into conventional scouting practice for gummy stem blight in Watermelon. Plant Dis 103:1642–1650

Kamilaris A, Prenafeta-Boldú F (2018) A review of the use of convolutional neural networks in agriculture. J Agric Sci 156:312–322

Kaspersen K, Berge TW, Goldberg S et al (2010) Estimation of weed pressure in cereals using digital image analysis. In: 3rd precision crop protection conference, 19– 21 September 2010, Bonn, Germany

Kaur S, Pandey S, Goel S (2019) Plants Disease Identification and Classification Through Leaf Images: A Survey. Archives Comput Meth Eng 26:507–530

Kim J, Morrison JR (2014) On the concerted design and scheduling of multiple resources for persistent UAV operations. J Intell Robot Syst: Theory Appl 74:479–498

Laothawornkitkul J, Moore JP, Taylor JE, Possell M, Gibson TD, Hewitt CN, Paul ND (2008) Discrimination of plant volatile signatures by an electronic nose: a potential technology for plant pest and disease monitoring. Env Sci Techn 42:8433–8439

Larsolle A, Muhammed HH (2007) Measuring crop status using multivariate analysis of hyperspectral field reflectance with application to disease severity and plant density. Prec Agric 8:37–47

Lee WS, Slaughter DC, Giles DK (1999) Robotic weed control system for tomatoes. Precision Agric 1:95–113

Li N, Grift TE, Yuan T, Zhang C, Momin MA, Li W (2016). Image processing for crop/weed discrimination in fields with high weed pressure. In: 2016 ASABE international meeting. american society of agricultural and biological engineers, pp. 1–11

Li Z, Paul R, Ba Tis T, Saville AC, Hansel JC, Yu T, Ristaino JB, Wei Q (2019) Non-invasive plant disease diagnostics enabled by smartphone-based fingerprinting of leaf volatiles. Nat Plants 5:856–866

Liu L, Lee SH, Saunders C (2014) Development of a machine vision system for weed detection during both off-sean and in-season in broadacre no-tillage cropping lands. Am J Agric Biol Sci 9:174–193

Lottes P, Hörferlin M, Sander S, Stachniss C (2016) Effective vision based classification for separating sugar beets and weeds for precision farming. J Field Robot 34:1160–1178

Lowenberg-DeBoer J, Huang IY, Grigoriadis V, Blackmore S (2020) Economics of robots and automation in field crop production. Prec Agric 21:278–299

Lu Y, Yi S, Zeng N, Liu Y, Zhang Y (2017) Identification of rice diseases using deep convolutional neural networks. Neurocomputing 267:378–384

Mahlein AK (2016) Plant disease detection by imaging sensors-Parallels and specific demands for precision agriculture and plant phenotyping. Plant Dis 100:241–254

Mahlein AK, Steiner U, Dehne HW, Oerke EC (2010) Spectral signatures of sugar beet leaves for the detection and differentiation of diseases. Precis Agric 11:413–431

Malneršič A, Dular M, Širok B, Oberti R, Hočevar M (2016) Close-range air-assisted precision spot-spraying for robotic applications: aerodynamics and spray coverage analysis. Biosyst Eng 146:216–226

Martinelli F, Scalenghe R, Davino S, Panno S, Scuderi G, Ruisi P, Villa P, Stroppiana D, Boschetti M, Goulart LR (2015) Advanced methods of plant disease detection. A review. Agron Sustain Dev, Springer Verlag/EDP Sciences/INRA 35:1–25

Menendez-Aponte P, Garcia C, Freese D, Defterli S, Xu Y (2016) Software and hardware architectures in cooperative aerial and ground robots for agricultural disease detection. In: Proceedings of the International conference on collaboration technologies and systems, Orlando, FL, USA, 2016, pp 354–358

Meroni M, Rossini M, Colombo R (2010) Characterization of leaf physiology using reflectance and fluorescence hyperspectral measurements. In: Maselli F, Menenti M, Brivio PA (eds) Optical observation of vegetation properties and characteristics. Research Signpost, Trivandrum, pp 165–187

Miller P, Lane A, Wheeler H (2000) Matching the application of fungicides to crop canopy characteristics. In: The BCPC 2000 conference: Pests and diseases, vol 2. British Crop Protection Council, Brighton, UK, pp 629–636

Mink R, Dutta A, Peteinatos GG, Sökefeld M, Engels JJ, Hahn M, Gerhards R (2018) Multi-temporal site-specific weed control of Cirsium arvense (L.) Scop. and Rumex crispus L. in Maize and Sugar Beet Using Unmanned Aerial Vehicle Based Mapping. Agriculture 8:65

Moshou D, Bravo C, Oberti R, West JS, Ramon H, Vougioukas S, Bochtis D (2011) Intelligent multi-sensor system for the detection and treatment of fungal diseases in arable crops. Biosyst Eng 108:311–321

Mutka AM, Bart RS (2015) Image-based phenotyping of plant disease symptoms. Front Plant Sci 5: art. no. 734

Naidu RA, Perry EM, Pierce FJ, Mekuria T (2009) The potential of spectral reflectance technique for the detection of Grapevine leafroll-associated virus-3 in two red-berried wine grape cultivars. Comput Electron Agric 66:38–45

Neto JC, Meyer GE, Jones DD, Samal AK (2006) Plant species identification using Elliptic Fourier leaf shape analysis. Comput Electron Agric 50:121–134

Nieuwenhuizen A, Hofstee J, van Henten E (2010) Performance evaluation of an automated detection and control system for volunteer potatoes in sugar beet fields. Biosys Eng 107:46–53

Nordmeyer H (2006) Patchy weed distribution and site-specific weed control in winter cereals. Prec Agric 7:219–231

Nutter FWJ, van Rij N, Eggenberger SK, Holah N (2010) Spatial and temporal dynamics of plant pathogens. In: Oerke EC et al (eds) Precision crop protection – the challenge and use of heterogeneity. Springer, Dordrecht, pp 27–50

Oberti R, Marchi M, Tirelli P, Calcante A, Iriti M, Borghese AN (2014) Automatic detection of powdery mildew on grapevine leaves by image analysis: Optimal view-angle range to increase the sensitivity. Comp Elect Agric 104:1–8

Oberti R, Marchi M, Tirelli P, Calcante A, Iriti M, Tona E et al (2016) Selective spraying of grapevines for disease control using a modular agricultural robot. Biosyst Eng 146:203–215

Oerke EC (2006) Crop losses to pests. J Agric Sci 144:31–43

Oerke EC, Steiner U (2010) Potential of digital thermography for disease control. In: Oerke EC et al (eds) Precision crop protection – the challenge and use of heterogeneity. Springer, Dordrecht, pp 167–182

Okamoto H, Murata T, Kataoka T, Hata SI (2007) Plant classification for weed detection using hyperspectral imaging with wavelet analysis. Weed Biol Manag 7:31–37

Olsen A, Konovalov DA, Philippa B et al (2019) DeepWeeds: a multiclass weed species image dataset for deep learning. Sci Rep 9:2058

Onyango CM, Marchant JA (2003) Segmentation of row crop plants from weeds using colour and morphology. Comput Electron Agric 39:141–155

Osterman A, Godesa T, Hočevar M, Sirok B, Stopar M (2013) Real-time positioning algorithm for variable-geometry air-assisted orchard sprayer. Comput Electron Agric 98:175–182

Paice M, Miller P, Bodle J (1995) An experimental sprayer for the spatially selective application of herbicides. J Agric Eng Res 60:107–116

Pantazi XE, Moshou D, Oberti R, West J, Mouazen AM, Bochtis D (2017) Detection of biotic and abiotic stresses in crops by using hierarchical self organizing classifiers. Prec Agric 18:383–393

Pedersen SM, Fountas S, Have H, Blackmore BS (2006) Agricultural robots—system analysis and economic feasibility. Prec Agric 7:295–308

Pérez-Bueno ML, Pineda M, Barón M (2019) Phenotyping plant responses to biotic stress by chlorophyll fluorescence imaging. Front Plant Sc 10:1135

Persson M, Astrand B (2008) Classification of crops and weeds extracted by active shape models. Biosyst Eng 100:484–497

Piron A, van der Heijden F, Destain MF (2011) Weed detection in 3D images. Precis Agric 12:607–622

Polder G, van der Heijden GWA, van Doorn J, Baltissen TAH (2014) Automatic detection of tulip breaking virus (TBV) intulip fields using machine vision. Biosyst Engin 117:35–42

Polder G, Blok PM, de Villiers HAC, van der Wolf JM, Kamp J (2019) Potato virus Y detection in seed potatoes using deep learning on hyperspectral images. Front Plant Sci 10: art. no. 209

Potena C, Nardi D, Pretto A (2017) Fast and accurate crop and weed identification with summarized train sets for precision agriculture. In: Advances in robot design and intelligent control. Springer International Publishing, pp 105–121

Pourreza A, Lee WS, Etxeberria E, Zhang Y (2016) Identification of Citrus Huanglongbing Disease at the pre-symptomatic stage using polarized imaging technique. IFAC Pap online 49:110–115

Raesch AR, Muller O, Pieruschka R, Rascher U (2014) Field observations with laser-induced fluorescence transient (LIFT) method in barley and sugar beet. Agriculture 4:159–169

Raja R, Slaughter DC, Fennimore SO, Nguyen TT, Vuong VL, Sinha N, Tourte L, Smith RF, Siemens MC (2019) Crop signalling: A novel crop recognition technique for robotic weed control. Biosyst Eng 187:278–291

Ramirez-Atencia C, Bello-Orgaz G, R-Moreno MD, Camacho D (2017) Solving complex multi-UAV mission planning problems using aponteapontemulti-objective genetic algorithms. Soft Computing 21:4883–4900

Riar DS, Ball DA, Yenish JP, Burke IC (2011) Light-activated, sensor-controlled sprayer provides effective postemergence control of broadleaf weeds in fallow. Weed Technol 25:447–453

Román C, Llorens J, Uribeetxebarria A, Sanz R, Planas S, Arnó J (2020) Spatially variable pesticide application in vineyards: Part II, field comparison of uniform and map-based variable dose treatments. Biosyst Eng 195:42–53

Rumpf T, Mahlein AK, Steiner U, Oerke EC, Dehne HW, Plümer L (2010) Early detection and classification of plant diseases with Support Vector Machines based on hyperspectral reflectance. Comp Electr Agric 74:91–99

Sankaran S, Mishra A, Ehsani R, Davis C (2010) A review of advanced techniques for detecting plant diseases. Comp Elect Agric 72:1–13

Sarri D, Lisci R, Rimediotti M, Vieri M (2014) RHEA airblast sprayer: calibration indexes of the airjet vector related to canopy and foliage characteristics. In: Proceedings of 2nd international conference on robotics and associated high-technologies and equipment for agriculture and forestry (RHEA-2014), pp 73–84

Sarri D, Martelloni L, Rimediotti M, Lisci R, Lombardo S, Vieri M (2019) Testing a multi-rotor unmanned aerial vehicle for spray application in high slope terraced vineyard. J Agric Eng L 853:38–47

Scholes JD, Rolfe SA (2009) Chlorophyll fluorescence imaging as tool for understanding the impact of fungal diseases on plant performance: a phenomics perspective. Funct Plant Biol 36:880–892

Schor N, Berman S, Dombrovsky A, Elad Y, Ignat T, Bechar A (2017) Development of a robotic detection system for greenhouse pepper plant diseases. Prec Agric 18:394–409

Šebela D, Olejníčková J, Sotolář R, Vrchotová N, Tříska J (2014) Towards optical detection of Plasmopara viticola infection in the field. J Plant Pathol 96:309–320

Slaughter DC, Giles DK, Fennimore SA, Smith RF (2008) Multispectral machine vision identification of lettuce and weed seedlings for automated weed control. Weed Technol 22:378–384

Slaughter DC, Giles DK, Tauzer C (1999) Precision offset spray system for roadway shoulder weed control. J Transp Eng 125:364–371

Sogaard HT (2005) Weed classification by active shape models. Biosyst Eng 91:271–281

Sogaard HT, Lund I (2007) Application accuracy of a machine vision-controlled robotic microdosing system. Biosyst Eng 96:315–322

Solanelles F, Escolà A, Planas S, Rosell J, Camp F, Gràcia F (2006) An electronic control system for pesticide application proportional to the canopy width of tree crops. Biosyst Eng 95:473–481

Su J, Liu C, Hu X, Xu X, Guo L, Chen WH (2019) Spatio-temporal monitoring of wheat yellow rust using UAV multispectral imagery. Comp Elec Agric 167: art. no. 105035

Swain KC, Nørremark M, Jørgensen RN, Midtiby HS, Green O (2011) Weed identification using an automated active shape matching (AASM) technique. Biosyst Eng 110:450–457

Teske ME, Wachspress DA, Thistle HW (2018) Prediction of Aerial Spray Release from UAVs. Trans ASABE 61:909–918

Tillett ND, Hague T, Miles SJ (2001) A field assessment of a potential method for weed and crop mapping on the basis of crop planting geometry. Comput Electron Agric 32:229–246

Tona E, Calcante A, Oberti R (2018) The profitability of precision spraying on specialty crops: a technical–economic analysis of protection equipment at increasing technological levels. Precis Agric 19:606–629

Too EC, Yujian L, Njuki S, Yingchun L (2019) A comparative study of fine-tuning deep learning models for plant disease identification. Comput Electron Agric 161:272–279

Torres-Sánchez J, López-Granados F, de Castro-Megías AI, Peña-Barragán JM (2013) Configuration and specifications of an unmanned aerial vehicle (UAV) for early site specific weed management. PLoS ONE 8:

Utstumo T, Urdal F, Brevik A, Dørum J, Netland J, Overskeid Ø, Berge TW, Gravdahl JT (2018) Robotic in-row weed control in vegetables. Comput Electron Agric 154:36–45

Van De Zande J, Achten V, Schepers H, Van Der Lans A, Michielsen J, Stallinga H, et al (2009) Plant-specific and canopy density spraying to control fungal diseases in bed-grown crops. In: Proceedings of the 7th European conference on precision agriculture, ECPA 2009. Wageningen Academic Publishers, pp 715–722

Vieri M, Lisci R, Rimediotti M, Sarri D (2013) The RHEA-project robot for tree crops pesticide application. J Agric Eng XLIV(s1):359–362

Vrindts E, De Baerdemeaeker J, Ramon H (2002) Weed detection using canopy reflection. Precis Agric 3(1):63–80

Waggoner PE, Aylor DE (2000) Epidemiology: a science of patterns. Annu Rev Phytopathol 38:71–94

Wahabzada M, Mahlein AK, Bauckhage C, Steiner U, Oerke EC, Kersting K (2015) Metro maps of plant disease dynamics-automated mining of differences using hyperspectral images. PLoS ONE 10:

Wang A, Zhang W, Wei X (2019a) A review on weed detection using ground-based machine vision and image processing techniques. Comp Elect Agric 158:226–240

Wang D, Vinson R, Holmes M, Seibel G, Bechar A, Nof S, Tao Y (2019b) Early detection of Tomato Spotted Wilt Virus by hyperspectral imaging and Outlier Removal Auxiliary Classifier Generative Adversarial Nets (OR-AC-GAN) Sci Rep 9: art. no. 4377

Wen S, Han J, Ning Z, Lan Y, Yin X, Zhang J, Ge Y (2019) Numerical analysis and validation of spray distributions disturbed by quad-rotor drone wake at different flight speeds. Comput Electron Agric 166, art.105036

Wen Y, Zhang R, Chen L, Huang Y, Yi T, Xu G, Li L, Hewitt AJ (2019b) A new spray deposition pattern measurement system based on spectral analysis of a fluorescent tracer. Comput Electron Agric 160:14–22

West JS, Bravo C, Oberti R, Lemaire D, Moshou D, McCartney HA (2003) The potential of optical canopy measurement for targeted control of field crop diseases. Annu Rev Phytopathol 41:593–614

West JS, Bravo C, Oberti R, Moshou D, Ramon H, McCartney HA (2010) Detection of fungal diseases optically and pathogen inoculum by air sampling. In: Oerke EC et al (eds) Precision crop protection – the challenge and use of heterogeneity. Springer, Dordrecht, pp 135–150

West JS, Kimber RBE (2015) Innovations in air sampling to detect plant pathogens. Annals Appl Biol 166:4–17

Wiesner-Hanks T, Stewart EL, Kaczmar N, Dechant C, Wu H, Nelson RJ, Lipson H, Gore MA (2018) Image set for deep learning: field images of maize annotated with disease symptoms. BMC Res Notes 11, art. no. 440

Woebbecke DM, Meyer GE, Von Bargen K, Mortensen DA (1995) Color indices for weed identification under various soil, residue, and lighting conditions. Trans ASAE 38:259–269

Wright DP, Baldwin BC, Shepard MC, Scholes JD (1995) Source-sink relationship in wheat leaves infected with powdery mildew. 1. Alterations in carbohydrate metabolism. Physiol Mol Plant Pathol 47:237–253

Xue X, Lan Y, Sun Z, Chang C, Hoffmann WC (2016) Develop an unmanned aerial vehicle based automatic aerial spraying system. Comput Electron Agric 128:58–66

Yang S, Yang X, Mo J (2018) The application of unmanned aircraft systems to plant protection in China. Precis Agric 19:278–292

Yue J, Lei T, Li C, Zhu J (2012) The application of unmanned aerial vehicle remote sensing in quickly monitoring crop pests. Intell Autom Soft Comput 18:1043–1052

Zhang B, Tang Q, Chen LP, Xu M (2016) Numerical simulation of wake vortices of crop spraying aircraft close to the ground. Biosyst Eng 145:52–64

Zhang Y, Slaughter DC, Staab ES (2012) Robust hyperspectral vision-based classification for multi-season weed mapping. J Photogramm Remote Sens 69:65–73

Zhang Y, Li Y, He Y, Liu F, Cen H, Fang H (2018) Near ground platform development to simulate UAV aerial spraying and its spraying test under different conditions. Comput Electron Agric 148:8–18

Zhou B, Wang J (2011) Discrimination of different types damage of rice plants by electronic nose. Biosyst Eng 109:250–257

Zhu H, Lan Y, Wu W, Hoffmann WC, Huang Y, Xue X, Liang J, Fritz B (2010) Development of a PWM precision spraying controller for unmanned aerial vehicles. J Bionic Eng 7:276–283

Chapter 7
Multi-robot Systems for Precision Agriculture

Angela Ribeiro and Jesus Conesa-Muñoz

Abstract This chapter addresses the integration of autonomous aerial inspection with autonomous ground intervention to perform agricultural tasks, specifically, precision treatments. Unmanned aerial vehicles (UAVs) may be used to inspect terrain and assess affected areas because of their suitability to access and cover large areas easily. Then, unmanned ground vehicles (UGVs) may use this information to perform crop treatments more efficiently and precisely. An overall system to manage these combined missions, comprising aerial inspections and ground interventions performed by a team of autonomous robots (multi robot system), was designed and implemented. Thus, two aerial robots were used to inspect the field autonomously, taking sets of images that were then processed to produce weed maps. From these weed maps plans were obtained for three autonomous ground robots that were coordinated to perform the treatments of the entire arable field. This chapter describes the coordination, planning and supervision or monitoring mechanisms implemented, as well as other interesting characteristics of the autonomous robot team.

Keywords UAV · UGV · Robot teams · Site-specific weed treatment · RHEA project

7.1 Introduction

In recent years, several research groups have examined the development of robotic technology to optimize the complex operations related to agriculture (García-Pérez et al. 2008; Auat Cheen and Caraelli 2013; Bechar and Vigneault 2016; Zhang et al. 2016) and in particular to precision treatment (Åstrand and Baerveldt 2002; Pedersen et al. 2006; Slaughter et al. 2008; Bakker et al. 2010). Most of the proposed approaches are based on single-robot systems, but there are important reasons to consider multi-robot systems as a more advantageous approach for agricultural tasks. In a generic way, Parker (2008) identified the following benefits, among others:

A. Ribeiro (✉) · J. Conesa-Muñoz
Centre for Automation and Robotics, CSIC-UPM, Arganda del Rey, Madrid, Spain
e-mail: angela.ribeiro@csic.es

© Springer Nature Switzerland AG 2021   151
A. Bechar (ed.), *Innovation in Agricultural Robotics for Precision Agriculture*,
Progress in Precision Agriculture,
https://doi.org/10.1007/978-3-030-77036-5_7

1. The task complexity is too great for a single robot to accomplish it.
2. The task is inherently spatially distributed.
3. It is often easier to build several robots with limited capabilities than a large complex robot.
4. The parallelism achieved through multi-robots accelerates task performance.
5. Redundancy in the multiple robot option increases robustness

In the context of agricultural tasks in general, and precision agriculture in particular, all of the above reasons are valid. Agricultural tasks tend to be distributed physically, and they are complex enough to take advantage of the use of more than one specialized machine. Furthermore, agricultural tasks can in general be focused using so-called mobile multi-robot systems, i.e. robotic systems that involve several mobile robots working as a team to perform well-defined tasks in clearly limited spaces with the aim of accomplishing a more comprehensive task, which will henceforth be called the *mission*, entrusted to the team.

The definition of a task for a single robot can become very complex since many aspects must be considered, including the static and dynamic characteristics of the robot. Management with a team of robots becomes even more complex, although the benefits of this type of team are evident, especially in agricultural tasks. In this case a solution based on a team of small or medium-sized robots is compared with one using a large conventional machine, even if it is equipped with many different advanced actuators and sensors. Table 7.1 shows some of these advantages.

An interesting type of multi-robot system is the heterogeneous multi-robot system characterized by the diversity of the robots, each contributing different capabilities. The heterogeneity may be evident as physical differences between robots or as behavioural differences when the robots fill diverse roles in a cooperating team. There are some examples of this type of system in the literature (Grocholsky et al. 2006; Viguria et al. 2012; Michael et al. 2012), which are often related to approaches

Table 7.1 Advantages of using a team of small or medium sized robots over one large agricultural vehicle

	A large machine or robot	A team of small or medium robots
Safety in autonomous operation mode	Becomes a safety problem in case of failure	Small or medium sized robots are more suitable to interact safely with humans
Fault impact on task completion	A failure will stop the entire task until the machine is repaired	Robot teams allow for re-planning of the task in the event of robot malfunction
Impact on the field	Considerable damage by soil compaction	Less compaction (lighter vehicles) and more precise movements (farming at plant level)
Personnel	An operator for each vehicle	One operator can supervise the entire robot team

that address some environmental problem. Although not integrated, many of these approaches comprise aerial robots for inspection tasks together with ground robots for intervention tasks to solve the problem. The proper integration of aerial and ground robots is especially needed in those contexts where an initial inspection of the affected zones allows better problem management later. This is the case for some agricultural tasks, such as selective treatments that can be observed as a global task split into two stages: an aerial inspection that can be performed by one or more autonomous aerial robots and a ground selective treatment that can be carried out by one or more autonomous ground robots properly equipped. Thus, the aerial inspection may provide a quick and easy assessment of the affected areas (for example, a weed map) to be used for ground intervention to implement more efficient treatment.

Multi-robot systems can be classified in several ways based on the motives behind their design. The multi-robot system proposed for site-specific treatments is focused on taking advantage of coordination between robots to improve the system performance. Consequently, the classification that best fits our goals is focused on coordination aspects. Farinelli et al. (2004) propose a taxonomy (Fig. 7.1) that considers two types of *dimensions* or specific features grouped together in the classification: *Coordination dimensions* to characterize the type of coordination and *System dimensions* referring to the system features that influence the development of the multi-robot system.

The *Cooperation level* distinguishes between cooperative systems, i.e. robots that operate together to perform a global task (Noreils 1993), and non-cooperative systems. Hereafter, we are interested in cooperative multi-robot systems because our proposal for site-specific treatment is based on such a system. The *Knowledge level* discriminates between systems comprising robots with some form of knowledge of

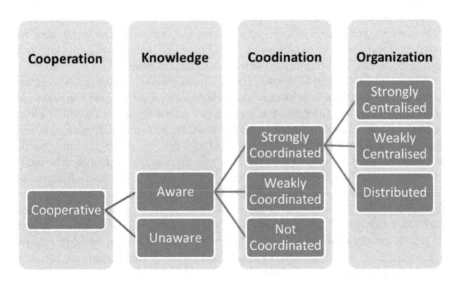

Fig. 7.1 Multi-robot system taxonomy. Coordination Dimensions

their teammates (*Aware*) and those where robots act without any knowledge of the other robots in the system (*Unaware*). Observe that knowledge and communication are not equivalent. Use of a communication mechanism does not entail awareness, and a multi-robot system can be aware even though there is no direct communication among the robots. The third level (*Coordination level*) addresses the cooperation mechanisms that allow the behaviour of the entire team of robots to be consistent though their actions, which are carried out considering the actions executed by other team components (Ferber 1999). The way that robots take into account the behaviour of their associates can be fixed by a set of rules known as a *coordination protocol*. The different degrees of coordination in the *Coordination level* depend on the use or not of the coordination protocol. Thus, *strong* coordination depends on a coordination protocol, whereas *weak* coordination does not. Weak coordination approaches are common in multi-robot systems because it is difficult to apply an effective coordination protocol in some instances. Finally, the *Organization level* is concerned with the way that decisions are made in the multi-robot system. The organization of a multi-robot system can vary from a fully centralized system, where a leader oversees organization of the work of the other robots, to an entirely distributed system, where the team robots are completely autonomous in the decision process and there is no leader.

System dimensions, in particular system features, such as communication, team composition, system architecture and team size, which are especially relevant in the system development must be taken into account in the design.

Communication among robots can be observed as a *cooperation* mechanism. Among the different types of communication available, direct and indirect communications are especially relevant for multi-robot systems. Direct communication is based on dedicated physical devices, whereas indirect communication shares information through modifications in the environment. The latter type of communication can reduce the complexity of large-scale system design as well as avoid the need for synchronization between the robots by providing a shared communication structure to be accessed by each robot in a distributed concurrent manner.

Team composition can be divided into two main classes: (i) homogeneous teams composed of members with exactly the same features, i.e. the same hardware and control software, and (ii) heterogeneous teams comprising robots that differ either in hardware or control software. In relation to the *system architecture*, different approaches can be considered such as *deliberative* or *reactive* architectures (Iocchi et al. 2000), and a hybrid proposal for an autonomous agrorobot can even be found in the literature (García-Pérez et al. 2008). A *deliberative* architecture for a multi-robot system enables the team members to cope with environmental changes by reorganizing overall behaviour of the team. In contrast, in a *reactive* architecture each robot manages the environmental changes by individually reorganizing its own task to accomplish the goal assigned to it. The main difference between the two approaches is the way unforeseen situations are tackled: in a deliberative architecture, the solution involves a long-term plan that considers all the available resources to accomplish a global goal collectively, in a reactive architecture, the plan affects only the robots involved in the problem. Finally, *team size* is an important issue that can be considered explicitly as a design choice, exploiting strategies to adapt team size to the complexity of the problem, as shown in the example described later.

A multi-robot system for an agricultural task is presented below, classifying it according to the previous taxonomy. To perform an agricultural task on a crop (covering several hectares), the robots of the team should always show cooperative behaviour. Although they do not need to know anything about the other robots in the team, it is necessary that an organization element (hereafter, the Mission Manager) be part of the multi-robot system; someone has to start the operation of the robots and send them the initial plans that they must carry out. Therefore, the multi-robot system is made aware, even though there is no direct communication among the robots. The proposed multi-robot system can be considered weakly coordinated; there is no need for a strong coordination protocol because once the robots receive their plans (commands), they execute them autonomously in disjoint spaces. In other words, by definition, the robots do not perform in the same space unless a malfunction occurs. Regarding organization, the multi-robot system for agricultural tasks is weakly centralized because during execution of the mission the robots perform their assigned tasks autonomously, even though the Mission Manager who has knowledge of the entire scenario generated the plans. In addition, task execution must always be supervised so that the operator knows what is happening at all times.

The communications are direct, but only between the robots and the Mission Manager. The team is heterogeneous if the integration of inspection (UAVs) and intervention (UGVs) is considered. The situations have also been considered (see below) where the aerial or ground robots themselves have different characteristics. The system architecture of the proposed system is hybrid, producing both deliberative and reactive behaviours. For example, when a robot stops working the system can re-plan the entire mission, taking into account all the available resources to accomplish a global goal collectively (deliberative), whereas in a potential collision among two or more robots, the system solution affects only the robots involved in the problem (reactive). Finally, the team size is fine-tuned by the planning strategies developed, which generate plans with minimum cost that may consider using fewer robots than those that are available in the team.

The remainder of this chapter describes the basic elements that compose a multi-robot system of inspection or intervention and the system developed to integrate and coordinate properly the aerial inspection and ground actuation. Both inspection and intervention will be considered hereafter as robot missions, and the software to integrate and coordinate all the components of such a multi-robot system is denoted as the Mission Manager. The proposed multi-robot system was initiated in the European project RHEA (Robot fleets for Highly Effective Agriculture and forestry management) as a heterogeneous multi-robot system composed of two small unmanned aerial vehicles (UAV) and three medium-sized ground vehicles (UGV) that were capable of cooperating with each other to automate site-specific treatments (Perez-Ruiz et al. 2015; Ribeiro et al. 2015; Gonzalez-de-Santos et al. 2016). The RHEA project was funded by the 7th EC Framework Programme, aimed to minimize the use of chemical treatments and reduce the time and energy needed for treatment, while guaranteeing the quality and safety of the products and making the implementation suitable for application to whole fields.

7.2   The Multi-robot System

The multi-robot system developed includes autonomous aerial and ground robots that have some minimal operational capabilities to allow the automation of the proposed tasks. The minimal set of operations or commands considered for both aerial and ground robots is summarized in Table 7.2; they match those of most market vehicles (Parrot Parrot 2016; Mota Commercial Drones 2016; Robotnik 2016; Clearpath Robotics 2016). Note that both the UGV and UAV missions can be performed effectively by selecting suitable sequences of these operations. In fact, the missions performed in the RHEA project were completed successfully using only that set of defined operations. The two types of missions considered were: (i) inspection missions where several UAVs acquired enough crop images to determine the spatial distribution of weeds and (ii) treatment missions where UGVs performed one of the following missions: weed spraying in cereals, mechanical and thermal weed control in maize, or the spraying of olive groves.

Both aerial and ground robots have sensors that enable them to acquire information on their internal state and the environment. This information is essential for supervising the task execution and keeping everything under control. Inertial measurement units (IMUs) and RTK-GPS receivers are common devices that are also integrated into the RHEA multi-robot system. The aerial robots must have additional sensors to perform the inspection. In fact, the aerial robots in RHEA are equipped, as are most current commercial UAVs, with visible and infrared cameras and GPS, enabling acquisition of geo-referenced images that can be processed later to extract relevant information. The ground robots must have suitable actuators to perform the task assigned, therefore, they can vary considerably. Sprayers, seeders and harvesting robotic arms are just a few examples (Blackmore et al. 2005).

Table 7.2 Operations considered for UAVs and UGVs

Operation or command	Description
Initialization	Robot configuration
Actuation	Robot action (displacement, speed change, tool activation, plan execution…)
Pause	Interrupt the current operation, keeping the state until a resume command arrives
Resume	Resume the operation that was being carried out prior to the pause command arrival
Stop	Stop the robot movement and actuation
Disconnect	Close the connection from which the request has been made

In the case of the RHEA system, all the information provided by the robots was transmitted for monitoring purposes to a Control Centre or Base Station, which was placed in a cabin next to the working area. The control station was equipped with powerful antennae, a router to create a wireless network to access the robots and a computer that gathered and processed all the data.

7.3 The Mission Manager

The Mission Manager, executed by the base station computer, integrates and automates both the inspection and treatment tasks. Its main goal is to automate the working sequence required by the multi-robot system. Although the robots are fully autonomous, high-level software is used to supervise execution of the task as well as to integrate the inspection results with the treatment. Thus, integration and coordination involves the generation and conveying of the proper trajectory to each ground robot for accomplishing the treatment, whilst the supervisor informs the operator about any unexpected behaviour detected during the task execution. Figure 7.2 shows the generic architecture of the Mission Manager and the interconnection with other

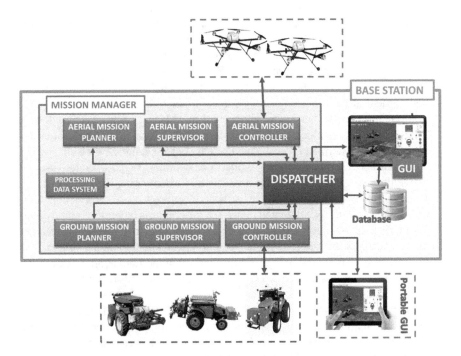

Fig. 7.2 Multi-robot system architecture. Main software modules (subsystems) and their interconnections

modules in the RHEA system. In the proposed architecture, we can distinguish the following modules (pieces of code):

- Aerial and ground mission planners. Several computer programs that generate plans autonomously for the two types of robots to complete their tasks because of the inherent differences between the aerial and ground robots as well as between inspection and treatment tasks. For example, the trajectory of a UGV is always constrained by its turning radius, whereas battery capacity is the main constraint of the aerial trajectories. The planners also have to take into account features of the sensors on board UAVs and UGVs when they affect the trajectories. For example, the image size and resolution determine the production of suitable trajectories for capturing high quality images that fully cover a certain area. Other important aspects should also be considered in generating the plan, such as fuel consumption, obstacles, travelled distance and robot speed to guarantee safe and optimal trajectories.

- Aerial and ground mission controllers. Units that combine hardware and software and are responsible for automating the task at the team level. They provide an interface to communicate with all robots for coordinating the team. Controllers enable team robots to engage in the simultaneous and cooperative execution of operations such as launch, pause, resume and stop. Moreover, the controllers are devoted to transmission of the plans generated to the robots and have to code them as an ordered set of commands that can be executed by the robots. In addition, the controllers must provide a stable channel to receive sensor information continuously from the robots so that control of the task is maintained.

- Aerial and ground mission supervisors. Several computer programs that provide high level monitoring, verify that a mission is executed according to the plan. The environment of the robots is usually subject to unpredictable conditions such as wind, changes in light, terrain roughness and animals that might cause small deviations from the scheduled plan. Supervisors receive relevant information about the robot from the controllers. Once deviations are detected, the supervision systems should, at least, send a notification (alarm or warning) to the operator (user) in charge of the mission. Deviations are detected by comparing the current state of the units (position, speed, and so on) provided by the on-board sensors (e.g. GPS, IMU) to the previously calculated plans. If a difference is larger than some predefined threshold, an alarm will be generated that reports the problem. Thresholds are set to avoid noise and false positives, with the aim of helping the human operator in charge of the mission to avoid missing any important detail. For example, the threshold margins used to detect anomalies in the trajectories and speeds were 30 cm and 1 km per hour, respectively.

- A data processing system receives and analyses the data acquired in the inspection task with the aim of extracting useful knowledge for the treatment task. In the RHEA case, this module is a mapping system that processes images taken by the aerial robots. It detects and obtains the coordinates of the weed patches to generate a weed distribution map that will be taken into account to produce ground robot plans.

- A dispatcher manages the entire workflow in the Mission Manager, integrating the inspection and treatment tasks. The dispatcher connects all the modules into the Mission Manager, redirecting processes to appropriate modules when required. Moreover, it manages external commands (plans, executions, pauses, resumes and aborts) when the operator (user) wants to control the workflow directly of the Mission Manager. The dispatcher allows the connection of new modules in such a way that the Mission Manager could have new functionalities.

In the proposed architecture for the whole RHEA system (see Fig. 7.2), in addition to the internal modules of the Mission Manager the following external modules are especially interesting:

- GUI (Graphical User Interface) displays the data generated by the Mission Manager, such as plans, execution states and alarms, guiding the operator through the different workflow steps. This module allows the operator to control the Mission Manager modules directly and consequently control the human–multi-robot system interaction. The main actions allowed are: planning aerial or ground tasks, launching aerial or ground tasks and processing the data acquired during the tasks. The GUI in RHEA was developed using the robot simulation environment Webots (Cyberbotics 2016). The interaction between the system and the human operator plays an important role because, in such a system, a vast number of events happen simultaneously and a human can easily miss critical information. To minimize this issue, the GUI was designed according to the strategy based on alarms proposed by Wilkins et al. (2003). The alerts are highlighted using small pop-up windows or eye-catching tags that report the most recent relevant events to the operator, e.g. the start of a new robot or even future events, such as collision warnings.
- Portable GUI allows the operator to control the robots individually outside the base station, i.e. in the field. The portable GUI is especially useful in case of breakdown.
- Database stores all data of the mission, e.g. plans, commands and telemetry. The stored data are required to interrupt and resume the process, or even to perform offline processes when the robots are not working, for example in the case of image processing or any other large data activity that could be important in the management of future incidents.

Because of their relevance for the integration and coordination of the multi-robot systems of inspection or intervention, the mission planners, controllers, supervisors and dispatcher are explained in detail in the following sections.

7.4 The Mission Planners

The planners focus mainly on collecting data related to the type of task (inspection or treatment) and field specifications (crop type, dimensions of fields, geographical position, and so on) given by the operator. With this information, the planner establishes the number of UAVs needed for inspecting the field as well as the number of UGVs needed to accomplish the treatment task. In both cases, it provides an action plan for each robot in the team.

The aerial planner splits the field into smaller rectangles optimally (the shape depends on the sensor of the camera), considering the orientation of the field, overlaps and resolution requirements for image analysis. The result of this process determines the exact positions (longitude, latitude and height) where the images have to be taken, and then it uses a Harmony Search algorithm to find the optimal order to cover them (Valente et al. 2013). This algorithm is a meta-heuristic optimization method that finds the shortest trajectory (global optimum), that is the order that produces the best harmony. The plans of the UAV are provided as an ordered list of points (waypoints) at which the cameras have to take images.

Similarly, the ground planner splits the field into parallel tracks and deduces the best sequence to cover them. Planning for the ground robots of the team is formulated as an kind of the Vehicle Routing Problem (VRP) and uses an approach (Conesa-Muñoz et al. 2016a; Conesa-Muñoz et al. 2016b) that integrates three different well-known meta-heuristic optimization methods (simulated annealing, the basic genetic algorithm and the non-sorting genetic algorithm II, also known as NSGA2) to find the best path for each ground robot and to determine the minimum number of robots required to cover the whole crop. The optimization approach can use different criteria simultaneously to minimize the distance travelled, the input cost and the time required to accomplish the treatment task, for example. The method developed also manages a team of UGVs with different capabilities and features (e.g. different turning radius and speed, among other factors). Figure 7.3 shows an example of a treatment task plan calculated for a team of three ground robots, taking into account a weed distribution map obtained after processing the data provided by the inspection mission. The weed distribution is represented in a matrix where components are labelled as weeds or non-weeds in the corresponding position of the field. In Fig. 7.3, the matrix components that contain weeds are bordered by black rectangles and must be sprayed by activating the nozzles at those points.

7.5 The Mission Controllers

The mission controllers are responsible for decomposing the mission plan into as many sub-plans as there are robots involved in the mission. A robot sub-plan comprises a sequence of commands that it understands. An example could be (i) open the connection with the robot, (ii) initialize the operation of the robot by setting

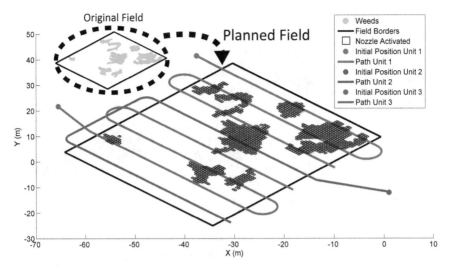

Fig. 7.3 Plan for spraying treatment of weeds with three ground robots

the origin of the coordinates, the base station location, and so on, (iii) send the sub-plan of the robot, (iv) verify successful reception of the plan, (v) wait until the mission ends, (vi) stop the robot, (vii) verify that the robot has actually stopped and (viii) close the connection with the robot. Therefore, the controllers are in charge of automating the execution of missions and of providing a high-level layer comprising basic operations of the robot to automate the cycle planning–execution–supervision. Note that the automation sequences of controllers depend on the events triggered by other modules, such as GUI requests, communication timeouts, robot confirmations and connection errors, as well as by previous events such as whether a similar request has been completed, whether the robot is already connected, or whether a response was received. All these, lead to the conclusion that mission controllers are reactive systems (Harel and Pnueli 1985; Iocchi et al. 2000) whose behaviours are difficult to predict because they react to asynchronous events based on their current states that are a consequence of the previously received stimuli. The behaviour of reactive systems is complex to analyse, and this makes them prone to failure. Thus, a careful design is essential, especially when potentially dangerous and expensive elements are involved, as in the case of the multi-robot system developed. The main problem in the design of this type of system is the formal and rigorous specification of the reactive behaviour. We have handled this difficulty by using state diagrams, such as those based on the *statechart* model (Harel 1987), taking advantage of the semantic richness of this model, which facilitates the development of comprehensible visual representations. Therefore, assuming the set of basic operations shown in Table 7.2, the state diagram for a generic mission controller, i.e. for both aerial and ground robots, is illustrated in Fig. 7.4. The 'H state' represents the last state within a super-state, so it may represent a running or paused state. The 'Init' is the

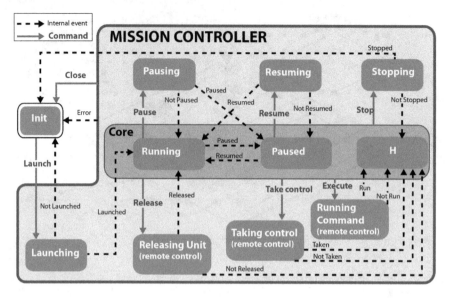

Fig. 7.4 State diagram of a mission controller

state that represents loading of the initial configuration of the mission controller as well as the close of its connection from which the request was made.

The high-level commands, shown as solid blue arrows in Fig. 7.4 are:

- Launch (a mission): The controller establishes connections with all the robots and sends a sub-plan to each one using the actuation commands shown in Table 7.2. Once a robot starts executing its sub-plan, the controller emits a *launched* signal, changing its state to *running*. If no robot starts, the controller emits the signal *not launched*, returning to its initial state.

- Pause (a mission): The controller sends a *pause* command to all robots, changing its state to *paused* after checking that all robots have fulfilled the order. Otherwise, the controller outputs the signal *not paused* and remains in the *running* state.

- Resume (a mission): The controller sends the *resume* command with the aim that the robots continue their execution after they have been stopped by a *pause* command. The state of the controller changes to the *running* state after at least one of the robots resumes execution.

- Stop (a mission): The controller sends a *stop* command to all robots. The mission ends when all robots perform the *stop* command, at which point the connections are closed, and the controller state returns to the initial state.

- Close (connections): This command closes all open connections when the robots have already stopped.

- Take control (of a robot for remote operation): This command lets the operator (user) take control of a robot on the team that is involved in a mission. The controller stores the remaining sub-plan before stopping the robot. Therefore, the robot is no longer considered part of the team that is executing the mission.

- Release (a robot): It returns the robot to the team. The robot resumes its sub-plan that was stored by the controller when the robot was taken off the team in favour of being controlled directly by the operator.
- Run Command (for remote operation): This command allows an operator to send commands directly (displacement, device setting and so on) to a robot that is no longer involved in the mission because it has been temporarily taken off the team by the take control command.

Most of the commands (launch, pause, resume, stop and close) act on the entire team, therefore, the controller must track the states of the robots with the aim of detecting internal signals or events that change the mission's state (dashed lines in Fig. 7.4). Note that many events come from external systems such as GUI requests or the paused or resumed signals received from lower-level controllers attached to the robots. However, there are also internal events caused by the mission controller that arise when a higher-level situation is detected. An example could be when all the robots have sent the paused (or resumed) confirmation; in that case, the mission controller assumes internally that the mission has paused (or resumed) and produces an internal signal to switch to the new state.

Some controller commands must be broken down into a sequence of lower-level commands. This is the case for the *launch (a mission)* and *release (a robot)* commands. In both cases, two requests are needed: first, robot initialization and second, sending of the mission sub-plan. The proper interpretation of these more complex commands requires the integration of a lower-level layer composed of as many controllers as there are robots in the team, called unit controllers for a mission. These controllers are directed to interpret the commands of the mission controller and to rewrite them using the command repertoire of the robot. Figure 7.5 shows the state diagram of one of these controllers. The main difference from the mission controller is that the internal signals, in dashed lines, are now independent of the team and depend on a single robot. In addition, the *launch* and *release* operations are split into several steps and expressed in terms of low-level operations that can be executed by the robots.

With the controllers proposed so far, the robots are able to confirm only the command reception but cannot guarantee their execution. Consequently, a new layer has been integrated into the architecture to provide a confirmation service for the command execution. The new controller (see Fig. 7.6) implements execution confirmation based on two facts: (i) the robots acknowledge the command reception (ACK signal) and (ii) the robots periodically convey some status information, such as their internal state (speeds, positions, and so on). This information comes from the sensor readings, and it is analysed by looking for occurrences of relevant events.

Every time the controller receives and processes a request, a related lower-level request is sent to the robot. Once the ACK message is received from the robot, the controller changes to a new state where it waits for some new status that supports execution of the requested operation by the robot. For detecting a change, the new status is reported by issuing an internal signal and switching to the new corresponding

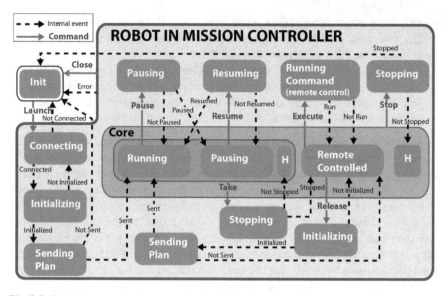

Fig. 7.5 State diagram of a robot controller for a mission

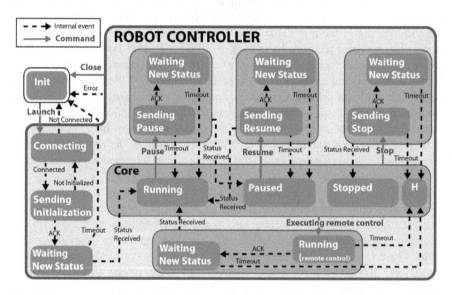

Fig. 7.6 State diagram of the basic controller of a robot

state. In contrast, if the ACK message is not received after a predefined time elapses, the controller assumes that the request was not executed and emits a timeout signal.

According to the descriptions so far, a generic high-level controller for missions is created as a multi-layer system to handle different levels of complexity. The proposed controller provides an execution confirmation service that enables management of a team of several robots with a precise response, i.e. knowing whether the command was received or not by the robots and whether it was executed successfully.

Note that individual robots can influence the overall mission controller. They send their sensor readings to the individual robot controllers, which process the information and react by changing their internal states (connecting, running, paused, stopped, waiting for something, and so on) and issuing the corresponding signals (connected, running, paused, resumed, stopped, and so on) when they detect the occurrence of these events.

These signals or events are the inputs for higher controllers that react similarly to them. At the end, they influence the global mission controller, changing their internal states.

Note that with the selected diagrams, the sequences of actions and transitions, but not the timing constraints, can be represented properly to complete a task in a given time, for example. With the proposed architecture, the supervision system takes care of these constraints because it runs during the mission, checking if tasks are being performed according to the expected schedule.

In the RHEA case, the UAV controllers are divided into two modules: a high-level module that decodes the plan generated by the planner and converts it into a sequence of commands supported by the robots and a low-level module that sends the commands and interacts individually with each aerial robot. The UGV controllers were implemented from the concepts outlined above using the state machine framework provided by the *Qt* libraries (Qt Libraries 2016). The three generic modules explained were adjusted to work with the ground robots of the RHEA team.

7.6 The Mission Supervisors

Both the aerial and ground supervisors analyse all the information received by the low-level controllers and check that the mission is performing according to the plans, reporting to the operator any unexpected situations by sending alarms to the GUI. The aerial supervisor is not very complex because, as a result of existing European legislation, UAVs must always fly under the close supervision of an expert pilot. For this reason, the aerial supervisor in RHEA can be regarded as a pilot support system that generates alarms such as a delay in mission execution, low battery level, or inaccurate trajectory. In contrast, the UGV can move autonomously in areas of several hectares, so the appropriate monitoring or supervision of the team of robots is essential. Consequently, the ground supervisor is more complex. The ground supervisor developed is a hierarchical structure comprising many simple supervisors that are can monitor the individual and collective behaviours of the different robots of the team.

To understand the proposed approach properly, it is important first to set out precisely some basic concepts such as mission, alarm and supervisor. The mission is so far the agricultural task that the robot team must perform and comprises a plan that contains, for each robot, the route, the speed and the tool state at each point of the route, for example in a spraying bar, the state (open or closed) of each nozzle. The alarm is the notification generated when some failure is detected, and at the higher level of supervision, it can also be a signal that some important events have occurred, such as the accomplishment of a route or the successful initiation of a device. Most of the alarms explained in this chapter are related to failures. Those cases where an alarm does not refer to malfunctions will be indicated explicitly. The supervisor is the module that periodically analyses the information received from the elements it monitors, i.e. engines, tanks, nozzles and sensors, or conceptual elements, such as routes and collisions. They generate an alarm when a failure is detected or an important event occurs. In the proposed approach, the supervisors are mainly composed of sets of IF–THEN rules that generate alarms if the information received meets particular rule preconditions. In general, the supervisor inputs can be expressed as a pair (property, value), where property indicates the element to be supervised and value indicates its current state. Supervisors produce several types of alarms. At the higher levels, they generate more types of alarms than at lower levels because they supervise elements that are more complex and therefore have to consider a more diverse set of failures and important events.

The proposed supervision architecture is distributed over different subsystems, taking advantage of the distributed nature of a team of tractors or robots working together on agricultural tasks. Supervision can be performed inside the robots themselves and can also be carried out by an external computer that monitors the work of the entire team and that is accessible to the operator. The supervisor levels are described in detail in the following sections.

7.6.1 Supervision Levels

The first level includes all the basic supervisors that run on the computers on board the robots. Thus, each basic supervisor is part of the Unit Control System (UCS) of each robot (see Fig. 7.7). Alarms that contain identification codes are generated when faults of the aboard subsystems are detected and when subsystems send alarms. In some situations, the faults can be solved by the subsystem itself or by the supervisor of the robot, in both cases without operator intervention. Alarms are always raised to higher levels for analysis under the perspective of the entire system, even if the detected failure is resolved inside the robot because low-risk alarms can be significant if they are combined with others. In summary, robot supervisors can detect and, in some cases, repair faults.

Ground units also send periodic monitoring messages to the second level, external to them, reporting on the robot's status. In the proposed architecture, the second level is divided into three main modules: (i) the Mission Supervisor, (ii) the Fault Recovery Module and (iii) the Alarm Notification Manager. The Mission Supervisor processes all data (alarms and monitoring messages) that come from the robots during mission execution, detecting more complex faults that involve more than one robot, more than one alarm or unexpected robot behaviour. This module also transmits the old and

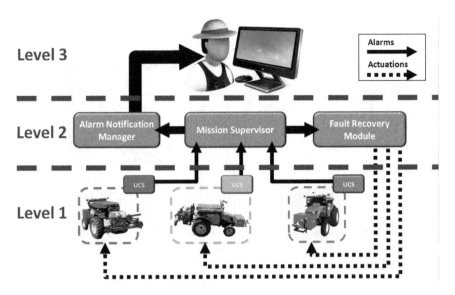

Fig. 7.7 Distributed multi-level supervision

new alarms to the Fault Recovery Module and to the Alarm Notification Manager. When the Fault Recovery Module receives an alarm, it uses a pre-established protocol to determine the action required to address the fault. Finally, the Alarm Notification Manager is a policy system that decides whether an alarm needs to be sent to the third level of supervision (operator), basing its decision on a set of policies that considers the priority and severity of the alarm.

The third level is used to convey information to the operator in charge of mission supervision. Therefore, this level is related to the Graphical User Interface (GUI). One of the goals of the proposed approach is that the operator receives sufficient information generated at the lower levels for the proper monitoring of mission performance. The operator (user) is the final decision element of the supervision architecture. If something does not work as expected, the operator can take control of the robot team and change the instructions for the mission execution. The Mission Supervisor as well as the Fault Recovery Module and Alarm Notification Manager are explained in detail in the following sections.

7.6.2 Mission Supervisor

The Mission Supervisor (see Fig. 7.8) comprises a set of individual supervisors working together to manage different items that are distributed across different levels: basic, robot and team levels. Thus, one supervisor monitors the speed of the robot, another monitors the route travelled, and so on. In this way, it is possible to execute only certain supervisors (if desired) or easily update one of them without affecting

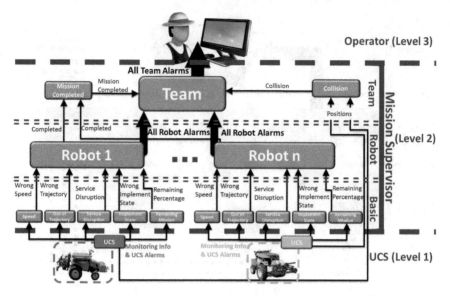

Fig. 7.8 Supervision architecture. The rounded rectangles represent the supervisors at different levels. Arrows represent the supervisor inputs or outputs, alarms and monitoring information

the others. Furthermore, by dividing and properly combining individual supervisors, more complex behaviour can emerge from the Mission Supervisor. The basic-level supervisors receive only data associated with a specific property that can be related to a physical entity, such as a nozzle or sensor, or a conceptual item, such as a route. Consequently, the basic supervisors contain only the logic necessary to detect failures related to their associated properties. At the robot level, the supervisors detect higher-level failures that arise from different properties of the same robot. For example, these supervisors can detect a fault condition in which more than one nozzle of the spraying bar does not work. Finally, at the team level, the supervisors detect anomalies pertaining to the behaviour of the entire team, for example, a collision involving several robots.

In addition to the information provided by the robots, the supervisors can also access mission data such as the defined routes for each robot, their speeds and tool states. These data do not change during the mission execution, so they are set for each supervisor at the beginning of the mission.

The internal logic of supervisor modules encapsulates fault detection of every supervised system. Furthermore, the alarms encapsulate the fault diagnosis because each type of alarm is associated with a type of failure.

In the proposed approach, the supervision behaviour is clearly decoupled because each supervisor encapsulates part of the supervision logic and the supervisors can all be easily replaced. Moreover, the proposed approach is hierarchical because it allows supervisors to link to each other, to form more complex supervisors that perform supervision at different levels. This combination of decoupling and hierarchy allows easy adaptation and updating of any supervision system, integrating in this way new supervision functions.

7.6.3   Fault Recovery Module

The Fault Recovery Module encapsulates fault recovery functionality, overseeing the repair of failures reported by the alarms. Thus, it receives the alarms issued by the Mission Supervisor and queries a specific database to identify the best strategy to resolve the alarm. The strategies comprise a set of actions that should be executed by the robot's on-board computer. For example, if a collision of several robots is predicted, the neutralizing strategy might be to avoid the collision by stopping all robots involved. In general, the strategies can concern one or more actions that must be executed by the computer on board the robots. Such actions might include reducing robot speed, changing the pressure of a nozzle, or restarting the mission.

7.6.4   Alarm Notification Manager

The Alarm Notification Manager decides when an alarm received should be sent to the operator (level 3) according to a set of predefined policies. Therefore, alarms can be filtered in certain situations, helping to generate alarms in a timely manner without overwhelming the operator with excessive, distracting messages. For example, consider a 'pilot flame' alarm in a mechanical-thermal tool that removes crop weeds (Raffaelli et al. 2013). In this type of implement, the pilot flame might be extinguished many times because of wind or other causes. The fault can be detected and solved by the implement itself because it can reignite the pilot flame repeatedly. However, if the alarm were permanently active, this larger problem can only be detected at higher levels of supervision, such as by the Mission Supervisor. It is clear that the situation described requires farmer or operator intervention to revise the tool operation, so he or she must be notified (level 3) by an alarm generated at level 2 related to the larger problem. In short, the operator need not be notified of the minor alarms generated by the on-board actuation system of the robot. For completeness, a log of all alarms generated during the mission is stored so that the operator can review it after the mission as required.

7.7   The Dispatcher

The dispatcher is the distributor module of the Mission Manager. It manages the interactions among the modules of the internal Mission Manager through a layer of abstraction that can replace any of the modules without having to make changes

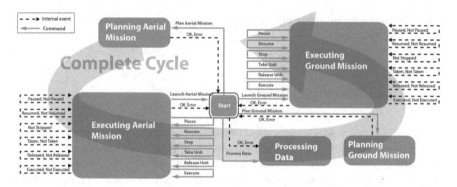

Fig. 7.9 State diagram for the dispatcher

in the interfaces. It redirects both GUI and portable device requests by generating the communication channel to the target module. Thus, the dispatcher arranges the human–system interactions. The interactions are needed to plan aerial or ground missions to launch missions and to process the data acquired during the mission performance. The workflow of the dispatcher is illustrated in the state diagram in Fig. 7.9.

Note that if the full cycle is executed, the aerial and ground missions are combined properly. In addition, the states associated with the planners and the data processing are set in such a way that the dispatcher invokes the aerial or ground planners or the associated data process, and then waits for their response when they end their execution. In contrast, the states associated with both the mission controllers and mission supervisors are set to allow all information provided by the controllers to be redirected to the supervisor for analysis.

7.8 Additional Aspects of the RHEA Multi-robot System

This section describes other RHEA system features that help to comprehend the complexity and magnitude of the multi-robot system developed.

The aerial robot team was made up of two six-rotor UAVs (AR200 model), developed by the AirRobot company (AirRobot 2016), each with a flight autonomy of approximately 40 min. Six-rotor units were used to provide a certain redundancy for safety in the case of failure in one motor. The aerial robots were able to provide telemetry information during the flight, including information required for supervision, such as estimating the position and battery level. The UAVs could carry a sensor payload of up to 1.5 kg. They were equipped with two cameras covering the visible and near-infrared spectra: two Sigma DP2 Merril models, with one modified to record NIR (near-infrared) images. The cameras were mounted on a gimbal system to reduce vibration and to allow them to point downwards when the UAV performs steady flights.

Georeferenced images of the crop were acquired and stored in the cameras' memory cards. The complete set of images was uploaded manually to the Base Station computer after the UAVs had landed. Then, a fully automated process produced weed maps of the entire field based on all of the acquired images. This process included a first stage to obtain an ortho-mosaiced image of the whole field, and a second stage to identify and locate the weed patches. The first stage was divided into the following three steps: (i) pre-positioning of the acquired images according to their time stamp and automatic detection, as well as labelling of artificial Ground Control Points previously placed around the field and georeferenced with an RTK-GPS receiver, (ii) registration of NIR and colour images using a specific algorithm based on Fourier analysis (Rabatel and Labbé 2014), (iii) mosaicing and geo-referencing of the final 4-band image (R, G, B, NIR) based on the MicMac (IGN, France) open-source photogrammetric software suite. Images with a 1-cm pixel resolution were obtained with a 60-m flight elevation. The residual error, including NIR-colour registration, was less than 0.5 pixels. Once the ortho-mosaiced image was generated, the weed map was obtained by an automatic and robust object-based image analysis (OBIA) procedure developed for mapping weed seedling patches in the very early pheno-logical stages: 2–4 true leaves (Peña et al. 2013). In addition, the OBIA algorithm was adapted to create a grid framework with a user-configurable size by applying a chessboard segmentation process. For example, according to the form of the sprayer boom developed in the RHEA project (a bar of 6-m length with 12 nozzles.), the grid size was 0.5×0.5 m.

The ground fleet was made up of three New Holland tractors of 50 hp (37.3 kW) and 1270 kg (New Holland 2016). Each tractor was adapted by reducing the driving cabin size and integrating the equipment required for perception, actuation, location, communication and safety on board the vehicle (Emmi et al. 2014).

An RTK-GPS receiver system, an RGB camera and a LiDAR sensor were integrated to ensure autonomous and safe navigation. The RTK-GPS receiver, a Trimble BX982 model, was a multi-channel, multi-frequency OEM GNSS receiver that provided centimetre-level positioning to the navigation system. The receiver supported two antennae, which enabled the heading of the vehicle to be estimated accurately. Therefore, a single connection to the tractor receiver (via RS232, USB, Ethernet or CAN) provided both centimetre-accuracy positions and a heading that was accurate to less than a tenth of a degree (2-m baseline between the antennae). In this way, both the positions and headings of the vehicles were provided with good precision at a frequency up to 20 Hz.

The camera on board each UGV was an SVS4050CFLGEA model from SVS-VISTEK (Seefeld, Germany) with a CCD Kodak KAI 04050 M/C sensor and a GR Bayer colour filter, which provided high-resolution images, i.e. 2336 by 1752 pixels with a 5.5-μm pixel size, to locate weeds, obstacles and crop rows accurately in real time (Ribeiro et al. 2005; Burgos-Artizzu et al. 2011; Guerrero et al. 2013). The camera was placed in a housing unit with a fan controlled by a thermostat for cooling, which allowed it to work even when raining or when the temperature was above 50 °C.

A LiDAR sensor, an LMS 111 (SICK AG, Waldkirch, Germany), was integrated into the UGVs to detect obstacles along the vehicle's trajectory with a ground clearance of 70 cm. The sensor was installed in the middle of the front of the vehicle with an inclination of 4 degrees, an angle obtained by trial and error.

To perform the treatment, the UGVs were equipped with different actuation equipment that includes physical (mechanical and thermal) or chemical (sprayer boom and air-blast sprayer) tools to destroy weeds (Frasconi et al. 2014) or apply pesticides (Perez-Ruiz et al. 2015). The selective sprayer boom comprised a 6-m length bar with 12 nozzles, that could be activated independently, and two tanks, one to store water (approximately 200 l) and another smaller one to hold herbicide. The sprayer was equipped with a direct injection system for mixing agrochemicals and water just before opening a single or several nozzles (Carballido et al. 2012).

The computer on board the UGV that executed the internal control system for managing the sensors and actuators, as well as for allowing remote control, was a CompactRIO model 9082 from National Instruments.

The computer at the Base Station was a desktop computer with an ASUS Z87-K SK1150/PCX 3.0 motherboard, an Intel core i7 4771 3.5 GHZ CPU, 2 DDR3 1600 8 GB PC3-12800 modules (16 GB RAM) and an SSD with 240 GB. The system was powerful enough to allow the simultaneous operation of all the active Mission Manager modules when performing a mission, including the controllers, the supervisors and the GUI.

The Base Station was also equipped with antennae and a router to create a multi-technology wireless network. A network based on a single technology was discarded because ISM (Industrial, Scientific and Medical) radio bands can be used without licensing, which might cause congestion and interference. Therefore, its performance might be impaired by devices that are not part of the communication system. The rigid communication requirements in terms of latency, throughput and connectivity that robotic control applications typically have are not guaranteed. Consequently, a wireless QoS-enabled multi-technology network based on the simultaneous use of multiple communication technologies (IEEE 802.11a, IEEE 802.11 g, ZigBee PRO and GPRS) was adopted to improve robustness and performance of the network (Hinterhofer and Tomic 2011). This reduces the risk that unpredictability of the wireless communication channel will disrupt the overlay communication because it is unlikely that transmission over several technologies will be affected at the same time. Furthermore, violations of the deadline caused by the congestion of the wireless channel can be reduced if the selection of communication technologies is coordinated among neighbouring devices.

7.9   Conclusions

A complex precision field task, i.e. composed of many different steps, can be tackled with a multi-robot system that properly combines a heterogeneous, aerial and ground, team and a Mission Manager that allows a single operator to supervise the entire process and manage the workflow required to complete the entire task or mission autonomously. This kind of multi-robot system allows the automatic sequencing of all the steps by using robots that support the usual minimal operations for many general tasks in many environments. For example, in agriculture, where there is currently no system able to link inspection data automatically to treatment actions.

A multi-robot system of this kind was developed in the European RHEA project and tested in different scenarios of precision treatment that include site-specific weed treatments. The inspection mission was used to acquire the data for generating weed maps that allowed the autonomous treatment of only the infested areas and consequently reduced both the cost of the treatment and the environmental impact. The success of the multi-robot system developed in RHEA introduces a new concept of applying a team comprising aerial and ground small or medium autonomous robots that effectively collaborate in the precise treatments of crops.

A spatially distributed and multilevel supervision is well-suited to the nature of a multi-robot system for agricultural complex tasks. Thus, the lowest level of supervision operates inside the robots, solving the most urgent issues. The higher level, executed in an external computer, is in charge of the more complex supervision that involves the entire team, and as a consequence, it is receiving information continuously from all the robots to have a complete overview of the status of the team. Finally, the third level allows a human operator to monitor in real time the entire work of multi-robot system and take control if needed. A modular and hierarchical architecture is a useful framework for obtaining incrementally complex supervisor systems that can be easily extended with other functionalities by adding new low-level supervisors.

Acknowledgements The RHEA project was funded in part by the 7th Framework Programme of the European Union under Grant Agreement No 245986-2. The authors want to express recognition to the RHEA beneficiaries: CSIC (Spain), CogVis (Austria), FTW (Austria), Cyberbotics (Switzerland), University of Pisa (Italy), University Complutense of Madrid (Spain), Tropical (Greece), AGROSAP (Spain), Polytechnic University of Madrid (Spain), AirRobot (Germany), University of Florence (Italy), IRSTEA (France), CNH (Belgium), Bluebotics (Switzerland) and CM (Italy).

References

AirRobot Website (2016) AirRobot Company Description. http://www.airrobot.de/. Accessed 20 Oct 2016

Åstrand B, Baerveldt A (2002) An agricultural mobile robot with vision-based perception for mechanical weed control. Auton Robots 13(1):21–35

Auat Cheen F, Caraelli R (2013) Agricultural robotics: unmanned robotic service units in agricultural tasks. IEEE Ind Electron Mag 7(3):48–58

Blackmore S, Stout B, Wang M, Runov B (2005, June) Robotic agriculture–the future of agricultural mechanisation. In: Proceedings of the 5th European conference on precision agriculture, pp 621–628

Bakker T, van Asselt K, Bontsema J, Muller J, van Straten G (2010) Systematic design of an autonomous platform for robotic weeding. J Terrramech 47:63–73

Bechar A, Vigneault C (2016) Agricultural robots for field operations: concepts and components. Biosyst Eng 149:94–111

Burgos-Artizzu XP, Ribeiro A, Guijarro M, Pajares G (2011) Real-time image processing for crop/weed discrimination in maize fields. Comput Electron Agric 75(2):337–346

Carballido J, Perez-Ruiz M, Gliever C, Agüera J (2012) Design, development and lab evaluation of a weed control sprayer to be used in robotic systems. In: Proceedings of the first international conference on robotic and associated high-technologies and equipment for agriculture, Pisa, Italy, 19–20 September 2012, vol 1, pp 23–29

Clearpath Robotics (2016). http://www.clearpathrobotics.com/. Accessed 20 Oct 2016

Conesa-Muñoz J, Pajares G, Ribeiro A (2016a) Mix-opt: A new route operator for optimal coverage path planning for a fleet in an agricultural environment. Expert Syst Appl 54:364–378

Conesa-Muñoz J, Bengochea-Guevara JM, Andujar D, Ribeiro A (2016b) Route planning for agricultural tasks: A general approach for fleets of autonomous vehicles in site-specific herbicide applications. Comput Electron Agric 127:204–220

Cyberbotics Website (2016) Webots Software. https://www.cyberbotics.com. Accessed 27 Oct 2016

Emmi L, Gonzalez-de-Soto M, Pajares G, Gonzalez-de-Santos P (2014) Integrating sensory/actuation systems in agricultural vehicles. Sensors 14(3):4014–4049

Farinelli A, Iocchi, L, Nardi D (2004) Multirobot systems: a classification focused on coordination. IEEE Trans Syst Man Cybern Part B (Cybernetics) 34(5):2015–2028

Frasconi C, Martelloni L, Fontanelli M, Raffaelli M, Emmi L, Pirchio M, Peruzzi A (2014) Design and full realization of physical weed control (PWC) automated machine within the RHEA project. In: 2nd International conference on robotics and associated high-technologies and equipment for agriculture and forestry (RHEA 214), pp 3–11

Ferber J (1999) Multi-agent systems: an introduction to distributed artificial intelligence, vol 1. Addison-Wesley, Reading

García-Pérez L, García-Alegre MC, Ribeiro A, Guinea D (2008) An agent of behaviour architecture for unmanned control of a farming vehicle. Comput Electron Agric 60(1):39–48

Gonzalez-de-Santos P, Ribeiro A, Fernandez-Quintanilla C, Lopez-Granados F, Brandstoetter M, Tomic S, …, Perez-Ruiz M (2016) Fleets of robots for environmentally-safe pest control in agriculture. Precis Agric:1–41

Grocholsky B, Keller J, Kumar V, Pappas G (2006) Cooperative air and ground surveillance. IEEE Robot Autom Mag 13:16–25

Guerrero JM, Guijarro M, Montalvo M, Romeo J, Emmi L, Ribeiro A, Pajares G (2013) Automatic expert system based on images for accuracy crop row detection in maize fields. Expert Syst Appl 40(2):656–664

Harel D, Pnueli A (1985) On the development of reactive systems. Springer, Berlin/Heidelberg, Germany

Harel D (1987) Statecharts: a visual formalism for complex systems. Sci Comput Progr 8:231–274

Hinterhofer T, Tomic S (2011) Wireless QoS-enabled multi-technology communication for the RHEA robotic fleet. In: Proceedings of the RHEA-2011 robotics and associated high-technologies and equipment for agriculture, Pisa, Italy, 19–21 September 2011, pp 173–186

Iocchi L, Nardi D, Salerno M (2000) Reactivity and deliberation: a survey on multi-robot systems. In: Workshop on balancing reactivity and social deliberation in multi-agent systems. Springer, Berlin Heidelberg, pp. 9–32

Michael N, Shen S, Mohta K, Mulgaonkar Y, Kumar V, Nagatani K, Okada Y, Kiribayashi S, Otake K, Yoshida K et al (2012) Collaborative mapping of an earthquake-damaged building via ground and aerial robots. J Field Robot 29:832–841

Mota. Mota Commercial Drones (2016) https://www.mota.com/drone/. Accessed 22 Oct 2016

New Holland (2016) http://agriculture1.newholland.com/nar/en-us/equipment/products/tractors-tel ehandlers/boomer-3000-series. Accessed 30 Oct 2016

Noreils FR (1993) Toward a robot architecture integrating cooperation between mobile robots: application to indoor environment. Int J Robot Res 12(1):79–98

Parker LE (2008) Multiple mobile robot systems. In: Springer handbook of robotics. Springer, Berlin Heidelberg, pp 921–941

Pedersen SM, Fountas S, Have H, Blackmore B (2006) Agricultural robots-systems analysis and economic feasibility. Precis Agric 7(4):295–308

Peña JM, Torres-Sánchez J, de Castro AI, Kelly M, López-Granados F (2013) Weed mapping in early-season maize fields using object-based analysis of unmanned aerial vehicle (UAV) images. PLoS ONE 8:

Pérez-Ruiz M, Gonzalez-de-Santos P, Ribeiro A, Fernandez-Quintanilla C, Peruzzi A, Vieri … M, Agüera J (2015) Highlights and preliminary results for autonomous crop protection. Comput Electron Agric 110:150–161

Parrot. AR Drone 2.0 (2016). http://www.parrot.com. Accessed 24 Oct 2016

Qt Libraries. Title (2016). http://www.qt.io/developers/. Accessed 28 Oct 2016

Rabatel G, Labbé S (2014) Registration of visible and near infrared aerial images based on Fourier-Mellin Transform. In: 2nd International conference on robotics and associated high-technologies and equipment for agriculture and forestry (RHEA 214), pp 329–338

Raffaelli M, Martelloni L, Frasconi C, Fontanelli M, Peruzzi A (2013) Development of machines for flaming weed control on hard surfaces. Appl Eng Agric 29(5):663–673

Ribeiro A, Fernández-Quintanilla C, Barroso J, García-Alegre MC, Stafford JV (2005) Development of an image analysis system for estimation of weed pressure. Precis Agric 5:169–174

Ribeiro A, Fernandez-Quintanilla C, Dorado J, López-Granados F, Peña JM, Rabatel G, …, de Santos PG (2015) A fleet of aerial and ground robots: a scalable approach for autonomous site-specific herbicide application. In: Precision agriculture'15. Wageningen Academic Publishers, pp 153–159

Robotnik. Mobile Robotnik Models (2016) http://www.robotnik.eu/mobile-robots/. Accessed 8 Nov 2016

Slaughter DC, Giles DK, Downey D (2008) Autonomous robotic weed control systems: a review. Comput Electron Agric 61:63–78

Valente J, Del Cerro J, Barrientos A, Sanz D (2013) Aerial coverage optimization in precision agriculture management: A musical harmony inspired approach. Comput Electron Agric 99:153–159

Viguria A, Maza I, Ollero A (2012) Distributed service-based cooperation in aerial/ground robot teams applied to fire detection and extinguishing missions. Adv Robot 24:1–23

Wilkins DE, Lee TJ, Berry P (2003) Interactive execution monitoring of agent teams. J Artif Intell Res (JAIR) 18:217–261

Zhang C, Noguchi N, Yang L (2016) Leader–follower system using two robot tractors to improve work efficiency. Comput Electron Agric 121:269–281

Chapter 8
Emerging Directions of Precision Agriculture and Agricultural Robotics

Ashwin S. Nair, Shimon Y. Nof, and Avital Bechar

8.1 Introduction and Definitions

In this chapter, we aim to shed some light on the next steps in the evolution of Precision Agriculture (PA) and Agricultural Robotics Systems (ARS), and the technological factors that will drive this evolution. To that end, we summarize a variety of research projects that are at the frontiers of Precision Agriculture and Agricultural Robotics Systems that integrate these two areas.

Precision Agriculture, as stated and discussed earlier in this book, is a field in agriculture concentrating on selective decision making and planning based on the processing of detailed farm-timely information, knowledge and thoughtful expertise. Underpinning Precision Agriculture is the need to improve aspects of the future farm, such as crop profitability and affordability, farm productivity and long-term sustainability, and environmental benefit. Precision agriculture is designed to follow these aims by reducing, through technological means, the required amount of fertilizers and other chemicals, irrigation, fuel, manual work, and lease and crop insurance payments (e.g. Mulla 2013).

In complex systems and systems-of-systems, intelligent control techniques and systems are necessary for dynamic, real-time interpretation and guidance of the environment and the objects operating in it (Nof 2009). Many PA related projects have been undertaken that use the potential of technologies and concepts, such as Cloud computing, Internet of Things (IoT), Internet of Services (IoS), Cyber Physical System (CPS), robotic simulators with realistic motion simulations, cyber augmented

A. S. Nair · S. Y. Nof (✉)
PRISM Center and School of IE, Purdue University, West Lafayette, USA
e-mail: nof@purdue.edu

A. Bechar
ARO, Rishon LeZion, Israel

© Springer Nature Switzerland AG 2021
A. Bechar (ed.), *Innovation in Agricultural Robotics for Precision Agriculture*,
Progress in Precision Agriculture,
https://doi.org/10.1007/978-3-030-77036-5_8

collaborative control and Human–Robot Collaboration. Some of these emerging technologies are described in this chapter.

8.1.1 Why Is Precision Collaboration Essential in Precision Agriculture?

The concept of Precision Collaboration (Bechar et al. 2015) is the underlying aspect in all emerging trends in Precision Agriculture. Why? Because many, often highly dispersed and distributed agents and resources are integrated to enable and accomplish the goals of PA. The details of Collaborative Control Theory and Precision Collaboration will be expounded in Sects. 8.4 and 8.5. Two key aspects of Precision Collaboration are:

1. When *networks and systems of systems scale up, and the probability of inefficiencies, gaps of responsibility, errors and conflicts increase, precise interaction becomes crucial. Therefore, it is worth implementing Precision Collaboration methods and tools.*
2. *Augmentation by sensors and collaborative control theory (CCT) enable and enhance smart and precise coordination and collaboration beyond communication and processing, and as contributors to collaboration support systems, has been found in recent research and surveys to be an important and valuable emerging area.*

A few definitions are included below because they are used often in this chapter:

1. **Cloud computing**: An information technology paradigm that enables ubiquitous access to shared pools of configurable system resources and higher-level services that can be provisioned with minimal management effort, usually over the Internet.
2. **Internet of things (IoT)**: A system of interrelated computing devices, mechanical or digital machines, objects and people that are provided with unique identifiers and the ability to interact and transfer data over a network without requiring human-to-human or human-to-computer interaction.
3. **Internet of services (IoS)**: A technology that provides the network infrastructure to support a service-oriented ecosystem. A fundamental characteristic of the IoS is that services combine and integrate collaboratively the functionalities of other services. (Van der Mei et al. 2018).
4. **Cyber physical systems (CPS)**: CPSs are commonly defined as the systems that offer collaborative integration of computation, networking and physical processes (Khaitan and McCalley 2015). The US National Science Foundation states "In cyber-physical systems, physical and software components are deeply intertwined, each operating on different spatial and temporal scales, exhibiting multiple and distinct behavioral modalities, and interacting with each other in a myriad of ways that change with context."

5. **Service oriented architecture (SOA):** A service-oriented architecture is a collection of services that communicate with each other. The communication can involve either simple data transfers or could involve dynamic coordination and collaboration among two or more services that combine temporarily for required purposes and timely execution.

6. **e-Work**: e-Work is a collection of collaborative, computer-supported and communication-enabled e-Activities, e-Operations, e-Functions and e-Support systems that enables other e-Systems and e-Activities (Nof 2003). The **c-Work** is a more advanced e-Work, augmented for smart collaboration by cyber-physical models and techniques. The Cc-Work is the currently emerging Cyber-Collaborative Work, enabled by cyber-augmented (e.g. wearables, augmented and virtual reality) human–robot–machine work processes and systems (Nof 2019).

7. **e-Service:** e-Service is the provision of services over electronic networks such as Internet, intranets or extranets without its scope being limited to service orga-nizations, but rather encompassing all enterprises, even those that manufacture goods and which require the development and implementation of sound service practices over electronic networks (Nof et al. 2015). The **c-Service** is a more advanced e-Service, where cyber-augmented collaboration is enabled.

8.2   Cloud Computing and Physical Internet/IOT, IOS and CPS for ARS c-Work and c-Service for Precision Agriculture

Precision Agriculture is an innovative effort that combines agricultural with digital and data science technologies that increasingly include cyber technologies, in the context of what is defined as ARS: Agricultural Robotic Systems. Innovations that involve various implementations based on cloud computing and Internet of Things/Services (IoT/S) into Precision Agriculture are expected to emerge in the future, given the rapid advancement and benefits of these technologies.

Cloud computing based applications of Agriculture IoT Sensor Monitoring Network were reviewed by Mekala and Viswanathan (2017a, b). A simple IoT model for an agricultural problem is presented in this chapter. The problem addressed was that farmers in India lack sufficient knowledge of soil characteristics and environ-mental information because the number of testing laboratories available in the country is limited. Internet of things based agriculture was proposed as a solution to this problem. The four layered IoT architecture can be applied to Precision Agriculture.

Layered architectures are commonly applied for the design and standardization of complex systems. Similar to layered architectures in industry and services, the first (top) layer in agriculture-related applications serves as a user interface layer. With this layer farmers can make decisions regarding crop protection and optimizing food production outputs and food security. The second layer involves data compi-lation, classification, processing, monitoring, and decision analysis. The third layer involves network management which would include communication technologies, such as Gateway, RFID, GSM, Wifi, 3G, UMTS, Bluetooth Low Energy, Zigbee, and so on. The fourth layer is the information collection layer that contains all physical instruments, sensors, cameras, and so on. This study also compares and contrasts various available hardware technologies and their use in an agricultural IoT setup. According to this survey, challenges for implementing IoT in agriculture include design of Service-oriented Architecture (SOA), Decision Support Systems (DSS) capabilities, efficient data mining and analytics, and IoT maintenance costs. The study addresses challenges and provides an IoT agricultural framework. Light Fidelity (Li-Fi) technology was introduced and evaluated for fixed area structure topology. The cloud computing framework was used to facilitate remotely controlled processes to perform spraying, weeding, bird and animal scaring, vigilance, moisture sensing, and so on. The methodology included smart warehouse management, which includes temperature and humidity maintenance, and theft detection. It also included intelligent decision making based on accurate real-time field data for smart irrigation with smart control.

Wang et al. (2014) also explored the architecture of the Internet of Things in agriculture with heterogeneous sensor data and proposed a data management system involving cloud computing to enable an IoT in agriculture (Fig. 8.1). Their design is based on a two-tier storage structure of a distributed database with large scalability, named HBase. Their work also proposes a management mechanism for heterogeneous sensor data for IoT in agriculture based on cloud computing. It consists of a data unification module, abnormal data processing module and a two-layer architecture to

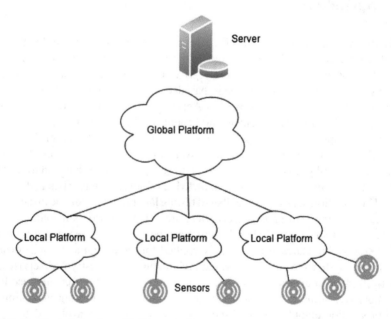

Fig. 8.1 Topological structure of IoT in agriculture

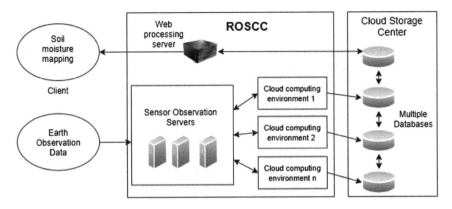

Fig. 8.2 Architecture of the ROSCC methodology

store data and access data. A cloud computing-based framework for agriculture information integration was also created by Duan (2012). In this research, a methodology and system for the integration of agricultural information and sharing a platform based on cloud computing were developed.

The data management problem of large size remote sensing images in soil moisture mapping for Precision Agriculture was addressed by Zhou et al. (2016). This methodology implements a Remote Sensing Observation Sharing method based on cloud computing (ROSCC) to enhance storage of remote sensing images and to achieve large-scale soil moisture mapping in Precision Agriculture (Fig. 8.2).

A system that combines wireless sensor networks (WSNs) and cloud computing into an integrated architecture for agricultural environmental applications was designed by Kassim and Harun (2017).

Cyber-Physical Systems (CPS) need to adapt to the changing physical world and expand their capabilities dynamically (Pradilla and Palau 2016). They designed a three-tier architecture that integrates: cloud computing, fog computing, and networks of sensors and actuators. The implementation involves the use of micro virtual machines and sensor observation, combining the isolation of virtual machines with standardized storage and information-exchange under a Sensor Web Enablement framework. The proposed architecture is coupled with the Internet of things (IoT) and applicable to Precision Agriculture.

The issue of information security and privacy in agriculture cloud information systems was addressed by Tan et al. (2014). Most encryption schemes cannot support encryption based on ciphertext. Therefore, it is difficult to build up the corporate and individual information security and privacy-securing in the information system based on a cloud computing platform. To enable information security and privacy of the cloud computing infrastructure that would be practical for an Agriculture Information System (AIS), the researchers have created an innovative encryption method for an agriculture intelligent information system (AIIS) based on a cloud computing platform. It is based on matrix theory and supports a series of cipher–text–operations? essential to create a secure communication protocol between users, owner and cloud server. This methodology can perform crypto-function at a moderate speed and can be used for securing corporate–individual privacy with regard to AIISs.

Further research will be needed to assure security of a cyber-augmented precision agricultural system and to prevent malicious disruptions and remote intervention in their safe and smooth operations.

8.3 Simulating an Agricultural Robotic System for Precision Agriculture Tasks

As more robotic systems are being developed and implemented in the agricultural domain, it would be cost effective to simulate such systems in the development phase. Recently there have been a few research projects on simulating a robotic system for human–robot collaboration. A computational simulation environment named 'Simulation Environment for Precision Agriculture Tasks using Robot Fleets' (SEARFS) was developed (Emmi et al. 2013) to study and evaluate the execution of agricultural tasks that can be performed by an autonomous fleet of robots. The environment is based on a mobile robot simulation tool that enables the performance, cooperation and interaction of a set of autonomous robots to be analysed while simulating the execution of specific actions on a three-dimensional (3-D) crop field. The SEARFS computational simulation environment is capable of simulating new technological advances such as GPS, GIS, automatic control, in-field and remote sensing, and mobile computing, which will enable the evaluation of new algorithms derived from PA techniques. This environment was designed as an open source computer application. The SEARFS environment consists of four levels of configurations, where the lower levels depend on the configuration of the higher levels (Fig. 8.3).

A general method for the development of customized robot simulation and control system software with a robot operating system (ROS) was also developed by Wang

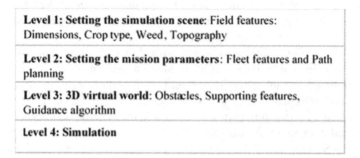

Fig. 8.3 SEARFS environment configuration levels

et al. (2016). The simulation designed in this research involves: a) a 3-D visualization model, created in URDF (unified robot description format) and viewed in Rviz to achieve motion planning with the MoveIt! software package, b) machine vision provided by a camera driver package in ROS to enable the use of tools for image processing, and 3-D point cloud analysis to reconstruct the environment to achieve accurate target locations and c) communication protocols provided by ROS for serial, Modbus support of the communication system development. A tomato harvesting scenario was simulated using this methodology to demonstrate its features and effectiveness.

8.4 Cyber Physical Systems and ARS

To overcome difficult problems such as the variability in agricultural produce and continuously changing conditions, development of intelligent systems is necessary to perform tasks successfully in such environments. Information acquisition systems, including sensors, fusion algorithms and data analysis need to be improved and adjusted to the dynamic and uncertain conditions of unstructured agricultural environments (Bechar 2010).

The trend in digital transformation has offered considerable opportunity for more efficient production using Cyber Physical Systems (CPS), which will enable new concepts for future farming systems (Herlitzius 2017). The rapid development of information and communication technologies is driving the evolution of mobile machines and devices into cyber-physical systems with few limitations with regard to communication.

A Precision Agriculture architecture (Fig. 8.4) was developed by Nie et al. (2014) based on CPS technology that comprises three control layers, i.e. the physical, network and decision layers.

A CPS oriented framework and workflow for agricultural greenhouse stress management, called MDR–CPS, was designed by Guo et al. (2018). It has been designed to focus on monitoring, detecting and responding to various types of stress. The system combines sensors, robots, humans and agricultural greenhouses as an integrated CPS, aimed at monitoring, detecting and responding to abnormal situations and conditions. The purpose is to provide an innovative solution that combines wireless sensor networks, agricultural robots and humans applying collaborative control theory (CCT) to detect and respond selectively to stresses as early as possible. The agricultural MDR–CPS framework is depicted in Fig. 8.5.

Sensor nodes are used in greenhouses to provide information on environmental properties that influence the healthy development of the agricultural crops. An agricultural cloud model platform is used in the field based on several server clusters (Guo et al. 2018; Zamora-Izquierdo et al. 2019).

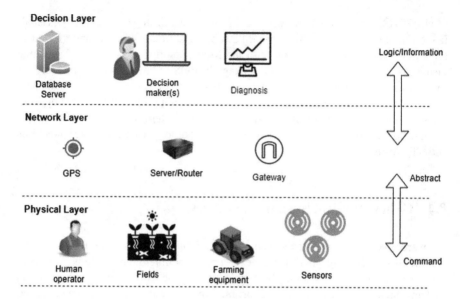

Fig. 8.4 Architecture of Precision Agriculture CPS nodes

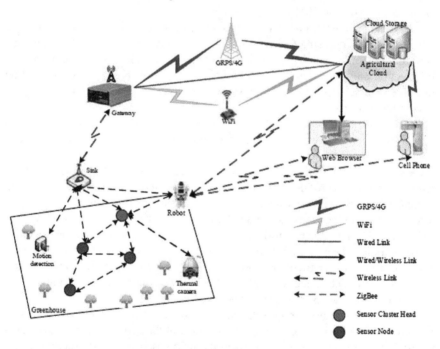

Fig. 8.5 Agricultural MDR–CPS framework (Courtesy of Guo et al. 2018)

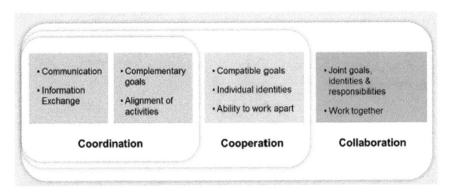

Fig. 8.6 Coordination vs Cooperation vs Collaboration in terms of interaction level (*Source* Nof et al. 2015)

8.5   Cyber-Augmented Collaborative Control of ARS

8.5.1   Collaborative Control Theory (CCT)

Collaborative control theory has been developed by researchers at the PRISM center at Purdue University and elsewhere (Nof 2007; Seok et al. 2012; Barbosa et al. 2014; Hernandez 2014; Nof et al. 2015; Yilmaz et al. 2017; Moghaddam and Nof 2017; Reyes Levalle 2018; Zhong and Nof 2020) to optimize distributed, decentralized and multi-agent based e-Work and s-Service. Collaboration is known to be essential for effective design and control of e-Work and e-Service. It enables all involved entities, human and artificial, in decentralized e-Systems to share their resources, information and responsibilities, such that mutual benefits are obtained (Figs. 8.6, 8.7 and 8.8).

Future precision agricultural systems will comprise multiple distributed and autonomous agents, therefore, the efficiency and effectiveness of the CPS would depend upon how well its constituent agents can collaborate.

Figure 8.9 illustrates the precision requirements for collaborative support features as evaluated by Bechar et al. (2015).

Automated processes in an uncertain and unstructured environment (such as agriculture) are challenged by changing peripheral requirements (Zhong et al. 2015). Addition of extra flexibility to the existing equipment to handle a larger range of tasks is a desirable solution, which can be offered, for example by Reconfigurable End-Effectors (REEs). An REE system has an adjustable structure to facilitate the adaptation of the end-effectors to various objects, therefore it is an intermediate solution between flexible and dedicated end-effectors (Zhong et al. 2015). Use of multiple end effectors enables the robot to adapt directly to multiple agricultural functions as and when required. For effective REE operations, the asynchronous cooperation requirement planning (ACRP) framework was created to facilitate the design and control of REE. The ACRP provides a dynamic solution, extending from the planning facet of collaborative control theory (CCT) for designing (offline) and

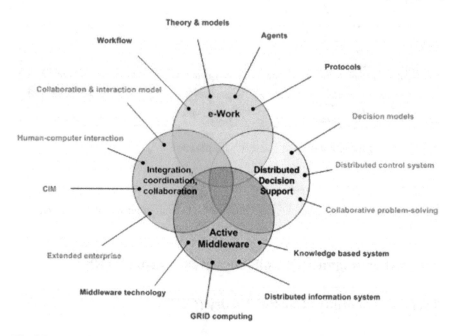

Fig. 8.7 The different components of cyber enhanced processes (*Source* Nof 2007)

controlling (online) multi-agent collaborations. The ACRP methodology is illustrated in Fig. 8.10.

The framework is illustrated with a case study of vegetable harvesting by multi-arm automated systems (Zhong et al. 2015). In harvesting processes, the grasp quality is one of the most important factors for production quality, therefore research on effective design and control of reconfigurable end effectors is highly relevant.

In emerging and future agricultural robotic systems, we can expect heterogeneity in multi-robot teams. To handle the varieties and variations of tasks observed in unstructured agricultural environments, multiple configurations of robots or heterogeneous robots, would need to be designed and included in the system. In such collaborative systems consisting of heterogeneous robots, ineffective task assignments can result in weak? collaboration and thus poor efficiency. Zhang et al. (2015) define the collaborative task assignment problem and develop a fuzzy collaborative intelligence-based algorithm to optimize the assignment plans as a solution to the challenging requirement of collaboration in heterogeneous multi-robot systems. This research introduces the concepts in collaboration type, the collaboration matrix and assignment matrix, and introduces an algorithm for adaptive fuzzy collaborative task assignment that is based on fuzzy set theory. Experimental results show and validate a shorter completion time, less energy consumption and a statistically significant larger loading accuracy. The methodology and algorithm were simulated in a general setting, and the methodology can be adapted directly to agricultural robotic systems.

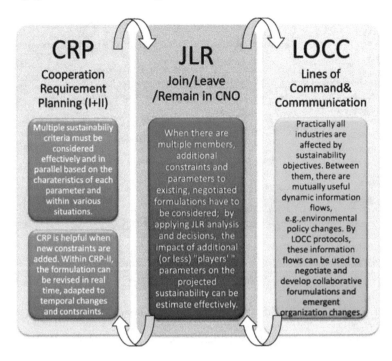

Fig. 8.8 Collaborative mechanisms of CCT and DSS (decision support system) for sustainability planning and control (*Source* Seok et al. 2012)

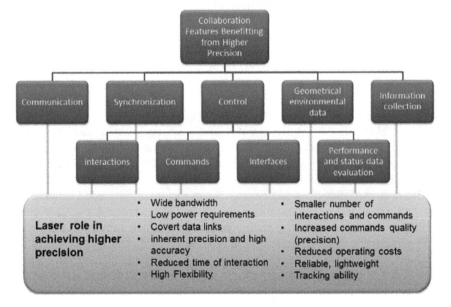

Fig. 8.9 Collaboration support augmented by laser: Features and their precision requirements (*Source* Bechar et al. 2015)

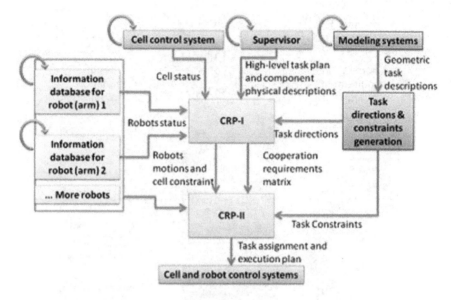

Fig. 8.10 Framework of Asynchronous Cooperation Requirement Planning (ACRP) (Courtesy of Zhong et al. 2015)

Below is a brief description of the types of collaboration in a multi-robot and a human–robot system (Zhang et al. 2015) (Table 8.1).

Research by Zhang et al. (2015) solves a heterogeneous multi-robot collaboration problem where stochastic and consecutive tasks are assigned to single or multiple robots in a dynamic changing environment.

The remainder of this section describes several recent research projects involving cyber enhanced and/or cyber augmented collaboration.

A. Methods for simultaneous orchard and harvesting robot design

Robotic manipulators can perform a variety of agricultural tasks, many of them with precision. However, despite decades of research, few agricultural robots have been commercialized. One of the reasons for the lack of agricultural robots on the market today is their high cost and lack of precision enabling functions, which makes them unprofitable for farmers.

Bloch et al. (2015, 2017, 2018) from the Agricultural Research Organization, Rishon LeZion, Israel prepared robotic systems that are optimal for specific tasks. In the optimization process, the robot's performance is maximized while allowing it to perform the task. To achieve a reliable result, the actual field task must be described and modelled with sufficient precision. However, the complex and unstructured environment of agricultural tasks complicates the task description as well as the robot-design process.

Table 8.1 Collaboration among robots working together (following Nof 1999)

	Collaboration types	Number of participating robots	Definition	PA task
(1)	Individual	One	One single robot completes task individually without any collaboration with other robots	Specific spraying, variable-rate applications, disease monitoring, etc.
(2)	Mandatory	Two or more	Two or more robots cooperate to complete a task simultaneously, and all of them are necessary for the completion	Cereal harvesting, fruit picking and storage, etc.
(3)	Concurrent	Two or more	Any one of the robots is able to complete the task, but when performed by two or more of them together concurrently, it decreases completion time, increases production or service quality and is more fault tolerant	Combined stress detection, yield assessment, complete plant protection system, etc.

The main goal was to characterize and analyse the environment of a given orchard and the required agricultural tasks, to understand their combined influence and interaction with the optimal design of a task-based robot for that orchard. This analysis allows the task description to be simplified by characteristics of the environment during simultaneous design of the robot and its environment.

The main results of the research are as follows. For the task-based robot optimization, we created a library with approximately 20 plant models. Software for evaluating the robot's performance effectiveness (optimization of cost function) was written and used for the optimal robot design. Based on the model library and software, robots were designed with optimal kinematics for a number of agricultural tasks and environments. During robot optimization, the level of complexity of the environment included yet did not enable the proposed software to solve the optimization problem in an acceptable time. In addition, a methodology for optimal robot location was developed.

To solve the robot-optimization problem for picking fruit in complex environments, a method was developed for characterizing the agricultural environment by fruit clustering and reaching cones. The method systematically reduces the complexity of the environments, thereby decreasing the number of calculations and providing a near-optimal solution. The method was approved and successfully applied to complex environments, solving the optimization problem in hours, rather than after weeks of calculations. The expected precision of the achieved solutions was 10% in the case examined.

A preliminary design for the robot working environment was prepared. Research findings include an environment that was fitted maximally to the robotic operation and that optimized one of the variables defining the structure of the environment.

In general, a set of tools and methodology was developed for analysis and design of the agricultural environment, together with optimal robot design. This methodology is novel in robot design, in particular in agriculture. It helps to improve the robot performance while designing low-cost robots affordable for farmers. The methods developed in this research are applied to apple and nectarine harvesting, although they can be used for robotic harvesting of any type of fruits, for other agricultural tasks, or in any area where the robot-environment design is used or is applicable.

B. Development of a selective autonomous sprayer for greenhouses

The essential process of pest control and chemical application of nutrients is one of the most important processes in any agricultural production. Nevertheless, the application requires human resources; it is a time-consuming task and exposes the operators to the danger of contamination with hazardous chemicals. Integrating autonomous robots and machinery for agricultural tasks involving expensive labour, and that are monotonous and hazardous has accelerated recently. An autonomous robot is an alternative in many cases. This research focuses on the development of a navigation procedure for an autonomous sprayer in a greenhouse growing sweet peppers.

C. A robotic sonar system for specific yield assessment and plant status evaluation

Specific yield assessment is essential for precision farming and agriculture in general. It is an important tool in agriculture for forecasting crop revenues, planning the budget and store capacity, labour management and compensation calculations (Fermont and Benson 2011). In several crops, such as fruit trees, fruit thinning is done based on yield estimation.

Subsidized crop insurance has become the most important single support policy in agriculture in both the USA and Israel. The program is immense in the USA, currently insuring over $120 billion in agricultural values and costing its taxpayers approximately $10 billion each year (Glauber 2013; Goodwin and Smith 2013). In Israel, for example, the aggregate premium payments from government subsidies for crop and disaster insurance programmes amount to over $25 million annually (source: Israeli Agriculture ministry budget).

An accurate and site-specific yield assessment technique that will decrease the assessment cost and increase its accuracy has the potential to reduce production costs, increase yield and profitability and save billions of dollars in tax subsidies:

The present techniques, however, for yield assessment are labour intensive and tend to be expensive. Moreover, the process is inaccurate because it is carried out manually by workers in the field and is based on crop sampling in small quantities, which in addition loses information on the variation. There is a tradeoff between the amount of time invested in sampling the crop and the accuracy given the inhomogeneous nature of crop distribution. To meet this challenge, various modern sensing technologies, such as thermal imaging (Stajnko et al. 2004), depth cameras (Andújar et al. 2016) and optical methods (Wachs et al. 2010) have been suggested for developing an automated system to detect crop biomass and for yield estimation (Lee et al. 2010). Recent vision-based studies in the context of specialty crops include one by Moonrinta et al. (2010) who developed a vision-based pineapple mapping algorithm with a detection success rate of 80%. Another vision-based yield estimation study by Nuske et al. (2011) detected 50–70% of the visible grapefruit and predicted the amount of crop mass with an error of 9%.

An ultrasonic sensing system was developed and the resulting classification features that would ultimately be used for a yield estimation robotic system were analysed (Mizrach et al. 2003; Mizrach 2008; Finkelstein et al. 2017). An algorithm was also developed to predict fruit mass per plant based on the ultrasonic echo return from a plant. The ultrasonic sensor system was tested in laboratory and greenhouse (with peppers) environments and on single pepper plants, single leaves and fruit. The results showed the potential of ultrasonic sensors for such a robot in classifying plants and greenhouse infrastructures such as walls. It showed the robot's ability to detect hidden plant rows and fruits as well as estimating the fruit mass in single plants. The system developed can detect and map crop rows without a direct line of sight using a matched filter and normalizing the acoustic energy by distance.

8.6 Human-Robot Collaborative System for ARS in PA Tasks

An overview and a framework for Precision Collaboration are shown below (Fig. 8.11). As mentioned in earlier sections, when networks and systems of systems scale up, and the probability of inefficiencies, errors and conflicts increases, the precision of interactions becomes crucial.

Unstructured environments such as agriculture are characterized by rapid changes in time and space (Bechar and Edan 2003). Fully automated systems do not perform well in such environments where they become cumbersome, complicated and expensive to develop and operate. Therefore, optimal output of a Precision Agriculture robotic system would depend on the effectiveness of collaboration between human agents and cyber controlled agents.

Cheein et al. (2015) reported a study that included guidelines for designing a human–robot interaction strategy for harvesting tasks that could be used for other agricultural tasks. The four design cores of a service unit are: mapping, navigation, sensing and action. This research addressed the problem of a decline in availability of human labour in agriculture in Chile and Argentina.

The research also discusses the current constrains related to precision farming and associated with flexible automation of farms in Argentina and Chile. The constraints include environmental constraints such as the variation in yield, the field, soil, crop, anomalous factors and management. For the latter this includes tillage practice and

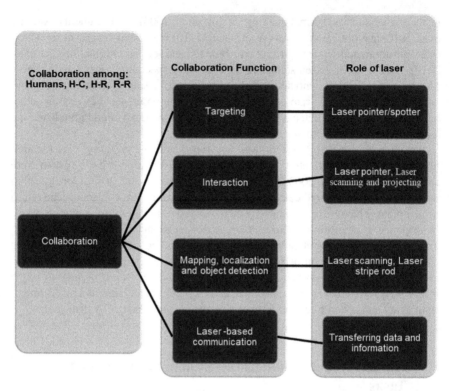

Fig. 8.11 Precision Collaboration support framework (H- Human; C- Computer; R- Robot)

seeding rate, crop rotation, fertilizer and pesticide application and irrigation pattern; these are facts that the service unit must know during its incursion into the workplace to avoid interference with the manual labour.

With regard to collaboration in a networked telerobotic environment, Nof and the PRISM center at Purdue University have developed a mechanism and tool called HUB-CI, a hub for Collaborative Intelligence (Fig. 8.12). Collaborative intelligence is a concept and a potential measure of performance of an e-System. It is a combination of communication intelligence, cumulative intelligence, cooperative intelligence and collective intelligence (Zhong et al. 2013). The HUB is an online portal that enables users to create and share research materials and computational tools. It can deliver all resources and simulations by a standard web browser and use high performance Grid computing resources. The majority of HUBs allow collaboration on virtual materials and simulations, but there has been no tool for users to perform physical collaboration (Zhong 2012). The HUB together with cloud computing allows software and data to be shared directly by groups of users, and provides knowledge and analytical tools that can be applied in Precision Agriculture systems (Nair et al. 2019; Sreeram and Nof 2021). The HUBs enable better, faster,

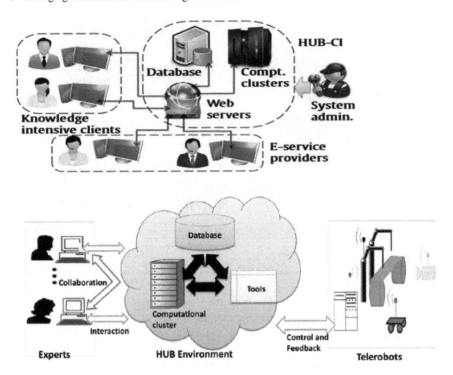

Fig. 8.12 The HUB-CI concept and architecture, enabling precision in operations planning and control (*Source* Zhong 2012; Devadasan et al. 2013)

smarter collaboration among decentralized, asynchronous decision-makers. Furthermore, it is considered a major enabler of precision in manufacturing, logistics and agriculture (Fig. 8.12).

The HUB-CI has been applied and tested in knowledge-based service planning (Zhong et al. 2013) and also to collaboration between telerobots and human agents in manufacturing (Zhong et al. 2013). Current research is being undertaken to apply HUB-CI in a telerobotic agricultural cyber physical system. Below are some projects in which Human–robot collaboration was applied to enhance output and productivity of the agricultural robot system:

A. Human–Robot collaborative system for selective tree pruning

Orchard pruning is a labour-intensive task that involves more than 25% of the labour costs. The main objectives of this task are to increase exposure to sunlight, control the tree shape and remove unwanted branches. In most orchards this task is done once a year and up to 20% of the branches are removed selectively.

A human–robot collaborative system for selective tree pruning has been developed (Bechar et al. 2014). The system consists of a Motoman manipulator, a colour camera, a single beam laser distance sensor, a human machine interface (HMI) and a cutting tool based on a circular saw developed for this task. The cutting tool, camera and

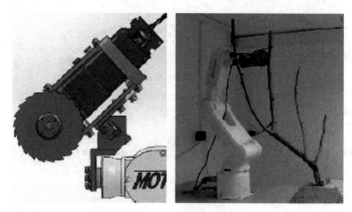

Fig. 8.13 Cutting tool design for tree pruning (*Source* Agricultural Research Organization Israel)

laser sensor are mounted on the manipulator's end-effector, and aligned parallel to one another (Fig. 8.13).

Experiments were established to examine the performance of the system under different conditions, human–robot collaboration methods and different trajectory types (Bechar et al. 2014). A cutting tool was designed for pruning branches with a diameter of up to 26 mm at a 45° cutting angle. The saw diameter was determined to be 115 mm with a standard shaft diameter of 41 mm. An interface to connect the cutting tool to the robot's end effector was designed to minimize the total dimensions of the tool and to increase e robot dexterity. An average cycle time of 9.2 s was achieved when the human operator and robot perform simultaneously. The results also revealed that the average time required to determine the location and orientation of the cut was 2.51 s.

B. Robot for automatic melon collection

Melon and watermelon harvesting require intensive manual labour. Machines with automatic robotic arms may replace personnel, especially in a simple routine that requires considerable physical effort. In this project a human is involved but in a different way. Based on preliminary tests it was found that about 80% of the workers's time is invested in transferring the picked melons from the bed and only 20% in locating and disconnecting the ripe melons from the plant. Therefore, the task is conducted in two steps. In the first, the human passes in the field, detects the ripe melons, marks their locations and disconnects them from the plants with pruning shears. In the second steps, the robotic system passes and collects only the melons that were marked and harvested. A robotic arm system has been developed (Fig. 8.14) that can collect the melons automatically knowing their coordinates, while moving through the collection area. An electro-mechanical robotic arm system has been assembled that consists of a wheeled frame, cylindrical rails with end limit switches, stepper motors with encoder for X- and Y-axis arm movement, a

Fig. 8.14 A close up of the melon picking robot and the robotic arm for melon picking (circled) (*Source* Agricultural Research Organization, Israel)

pneumatically operated robotic arm system for additional Y- and Z-axis movements, vacuum operated gripper, motor controllers and a PLC.

A human machine interface has been developed to enable operator intervention. A melon 'picking-up' simulator program has been created, capable of demonstrating the process of collecting melons by the robotic arm. For experimental applications, the melon collecting path optimization algorithm was used. The system was tested and succeeded in reaching up to seven target points in sequence with an accuracy of 84% (with a target reaching error of 7–10 mm, collection time 7–8 melons min^{-1}, at a distance of up to 4000 mm, with arm velocity of up to 800 mm s^{-1} and acceleration of up to 50 m per s^2).

C. Multi-sensor fault tolerant learning algorithm in an agricultural robotic system

Ajidarma 2017; Ajidarma and Nof 2021 aimed to develop a new fault tolerant interface design based on the collaborative control theory (CCT) principles of best matching (BM), error prevention and conflict resolution (EPCR) for an agricultural robotic system. They developed a fault tolerant learning algorithm to process the data of moving sensors in an agricultural robotic system. The sensor data and actual state of the object were modelled as a function of error and rate of conflict. Two learning algorithms, adaptive learning algorithm (ALA) and cumulative learning algorithm (CLA) were developed and tested. This method involves collaboration with a human operator and an adaptive learning mechanism to minimize measurement and detection errors. It is an excellent example of the concept of Precision Collaboration. This research addressed the problem of having an interface with fault tolerant sensor data processing in a collaborative agricultural robotic system where multiple sensors are mounted on a mobile robot, and a human operator performs supervisory functions.

D. Human–robot collaborative system for early detection of crop diseases

Traditional agricultural management practices assume that fields growing crops have homogeneous properties (Oerke et al. 2011). In contrast, modern, Precision Agriculture integrates different technologies, such as: sensors, information and management systems for adapting agricultural practices to variation within the field (McBratney et al. 2005; Dong et al. 2013). Monitoring is a major component in Precision Agriculture and of precision crop protection (Gebbers and Adamchuk 2010; Schellberg et al. 2008).

Crop yield is affected by different stresses, e.g. pests, diseases, weeds, environmental conditions, nutrition or water deficiencies, which can impair production. Oerke and Dehne (2004) indicated that the impact of diseases, insects and weeds represents a potential annual loss of 40% of world food production. The occurrence of diseases depends on environmental factors and they often have a sporadic spatial distribution, therefore sensing techniques can be useful in identifying primary disease foci and distribution (Franke and Menz 2007; Franke et al. 2009). Sankaran et al. (2010) and Lee et al. (2010) suggested that detection and quantification of diseases with visible and infrared spectroscopy would be feasible. If a symptom or a disease can be detected by the naked eye, a sensor should be able to record the stress symptoms (Nutter et al. 1990; Stafford 2000).

Currently, disease detection and monitoring in greenhouses are conducted manually by an expert inspector and are limited because of the availability of human resources, sparse sampling and large monitoring costs. Sampling intensity and resolution are low with about 20 arbitrarily locations sampled per hectare in a fixed pattern (the same locations are revisited) and each plot is monitored every 7–10 days. The plants are inspected for symptoms by an inspector crossing the greenhouse rows on foot. Thus, the inspector walks about 20 km per day covering about 8 hectares, and a designated inspector is required for every 80 hectares. The limitations of the current inspection methods can lead to late detection and inability to contain a disease. As a precaution, repeated, large doses of pesticide are often applied even when symptoms are far below thresholds that require pesticide application. Moreover, pesticides are typically applied uniformly throughout the greenhouse while disease distribution is typically centred in distinct locations, resulting in additional pesticides use, increased material cost and adverse environmental effects.

In greenhouses, a current challenge is the early detection of stresses (potentially leading to diseases) and other crop risks to prevent the spread of uncontrolled disease and hence improve productivity. Often detection is too late even though there is enough knowledge on how to address the specific stress in plants. Different biotic and abiotic stresses affect the expected potential crop yield. These stresses and other factors that limit yields must be detected as early as possible such that appropriate and precise counter measures may be applied. In the absence of an affordable and effective monitoring mechanism or system, the decisions taken by farmers could be wrong and might result in over- or under-application of pesticides, nutrients and water.

Robotic systems in greenhouses enable early detection and improved control of plant diseases. They are expected to increase yield, improve quality, reduce pesticide application, increase sustainability and reduce costs. Symptoms vary for each disease and crop, and each plant might suffer from multiple threats, thus, dedicated integrated disease detection systems and algorithms are required.

Automation of disease detection can alleviate current difficulties and lead to improvement in yield together with considerable reduction in pesticide use (Franke and Menz 2007; Franke et al. 2009; Bock et al. 2010). In addition to reduced production costs, this will also lead to reduced exposure to pesticides of farm workers and inspectors, and increased sustainability (Hillnhuetter and Mahlein 2008). Plant diseases can affect various optical foliage characteristics, therefore disease detection can be based on different spectral ranges (Lee et al. 2010). Image processing of foliage light reflection has been applied to many different diseases and cultivars (for reviews see: Barbedo and Garcia 2013; Pujari et al. 2015; Patil and Kumar 2011; Lee et al. 2010). Methods based on fluorescence (Wetterich et al. 2016) or thermography (Oerke et al. 2011) can also be used for disease detection and have been extensively studied, but they are less relevant for a robotic detection system operating in the field because of cost, payload weight or required setup. Mobile robotic manipulators with various sensing capabilities offer an automated solution suitable for disease detection in greenhouses. There has been, however, little comprehensive research on the development of such integrated robotic disease detection systems for greenhouses, probably because the primary challenge of developing robust disease detection algorithms is still an open research question. Aerial platforms (Gennaro et al. 2012) and ground mobile robotic platforms with fixed sensor configurations (Harper and McKerrow 2001; Moshou et al. 2011; Pilli et al. 2014) for disease detection have been tested for open field crops. Yet, in greenhouses both solutions have inherent shortcomings. The maneuverability and flight duration of aerial systems within greenhouses is limited, and navigation and location cannot rely on GPS sensors because the structure can cause unpredictable errors, therefore they lose their main outdoor advantage. In greenhouses, sensory position and adaptation of orientation can greatly improve detection, especially early detection where symptoms are typically centered on distinct locations. For fixed sensor configuration, position and orientation adaptation are not possible. Moreover, in fixed configuration systems, the requirement for multiple disease detection can lead to a requirement for multiple detection positions and orientations, which tend to increase system complexity and cost and hinder maneuverability.

To address this issue, a robotic disease detection system for greenhouse pepper plants was developed based on the concept of a mobile robotic manipulator (Schor et al. 2015; Schor et al. 2016), which provides the required maneuverability and flexibility (Fig. 8.15). Prior to the above, no major system had been developed for disease detection for specialty crops in greenhouses that involved a mobile robotic manipulator.

The robotic disease detection system was developed holistically, i.e. system architecture, operation cycle and detection algorithms for multiple threats to a pepper crop were developed in an integrated manner. Eizicovits et al. (2016) showed that early

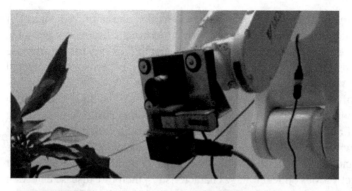

Fig. 8.15 The apparatus for disease detection for pepper plants (*Source* Agricultural Research Organization, Israel)

integration and testing of perceived requirements can lead to improved system design and operation in environments with taxing needs (e.g. the agricultural environment).

The detection system comprises a mechanical structure, sensor suite, motion planning (Fig. 8.16) and disease detection algorithms. Visual spectrum imagery is used for motion planning and disease detection for fast, non-destructive and cost-effective operation. An algorithm based on principal component analysis (PCA) was developed for powdery mildew, and three algorithms were developed for tomato spotted wilt virus)TSWV(disease detection, one based on PCA and two on the coefficient of variation (CV). Principal component analysis is a statistical tool used to reduce the

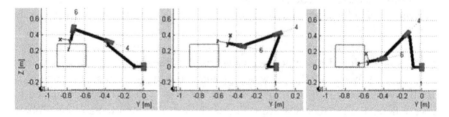

Fig. 8.16 Example of motion planning for the robotic arm in disease detection in plants (*Source* Agricultural Research Organization, Israel)

dimensionality of data and demonstrate patterns in a dataset. The CV is a statistical measure of dispersion, calculated as the ratio of the standard deviation to the mean.

The algorithms were tested using images of healthy and infected plants taken from a greenhouse. For RGB-based detection of TSWV, PCA-based classification with leaf veins removed achieved the greatest classification accuracy (90%), and the accuracy of CV methods was also high (85%, 87%). For powdery mildew (PM), the accuracy of pixel-level classification was high (95.2%) while that of leaf condition classification was low (64.3%) because leaf images came from the top of the leaf and disease symptoms start appearing on the bottom. The NIR-R-G-based detection produced inferior results for both diseases. The components of the system were integrated, and preliminary integration tests were done in a laboratory environment to verify that all system components would work together. The integrated system operated successfully for 110 consecutive minutes with an average cycle time of 26.7 s for end-effector velocity of 15 mm s^{-1} and PCA-based detection algorithms. Future research will examine improvement of disease detection, aiming to achieve greater accuracy together with earlier detection, e.g. by facilitating PM examination on the bottom of the leaf or by integration of the two CV-based methods. For complete integration tests and field performance studies, a dynamic detection process (i.e. with a moving conveyor) will be implemented and tested.

Results are encouraging because the cycle time attained was slower than the calculated required baseline (Schor et al. 2017). However, the laboratory environment comprising a conveyor belt, stationary sensor system and black background for simplifying plant identification and background removal procedures makes the disease detection task easier and faster. Conducting a disease detection task in an unstructured environment such as a greenhouse will require more sophisticated algorithms for motion control, path planning and image processing because of a more complex environment that includes obstacles, background noises, illumination etc., thus cycle time may be extended.

A subsequent multidisciplinary project was undertaken by researchers at the Agricultural Research Organization (ARO) in Israel, PRSIM center – Purdue University, USA and the University of Maryland, USA. This research was funded by BARD,[1] the US–Israel binational agricultural research and development fund (Bechar et al. 2020). It combines the following three disciplines to solve the problem of consistent early detection:

(1) Smart agricultural robots
(2) Human–robot collaboration (based on the HUB-CI and CCT described in Sects. 8.4 and 8.5)
(3) Early stress detection and classification using multispectral imaging and image classification and creation of a stress map

The robotic platform (cart) was modified at ARO to improve the control and autonomous navigation, and to suit the disease detection task better in a greenhouse. The platform is equipped with a UR5 manipulator, a sensory system comprising

---

[1] BARD Research Project IS-4886-16 R.

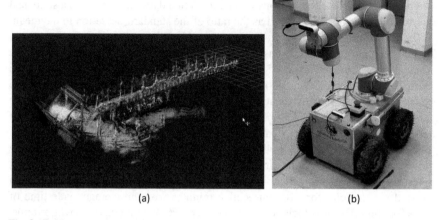

Fig. 8.17 Three-D mapping of a pepper greenhouse (**a**) and the robotic platform (**b**)

two depth cameras to create 3-D and 2-D maps of the greenhouse, the Kinect V2 and RealSense 435 and an RGB 1080p camera. A real-time environment mapping application was developed and modified with the robot sensors while it moves in the environment and generates a 3-D model of it. A 3-D mapping experiment was conducted in the laboratory and in a pepper greenhouse at ARO (Fig. 8.17).

For the 'human-in-the-loop' tasks of the agricultural robot system, a HUB-CI (hub for collaborative intelligence) system was developed by the PRISM team at Purdue and the ARO team. The objective: To enable effective and timely integration, and resulting collaboration tasks, by optimized exchange and leveraging of signals and information gathered in real-time from distributed components. The outcome of the HUB-CI is collaborative intelligence from the ARS networked components, thus enabling precision tasks (Nair et al. 2019). The following algorithms and protocols were developed by Dusadeerungsikul and Nof (2019): (a) algorithm to determine what image or case must be reviewed by remote human users, (b) adaptive search: use knowledge-based information, (c) routing algorithm: create a tour for a mobile robot, (d) detection-routing protocol: mechanism for remote disease detection algorithm to communicate with the routing algorithm, (e) manual control protocol: mechanism and constraints for manual control of the robot and (f) human-in-the-loop protocol: mechanism for human operator to communicate with the search and routing algorithm. The HUB-CI system has been designed as a virtual platform to integrate signals, data and control logic from several participating agents (cyber and human agents). It enables the cyber-collaborative protocols to make local control decisions based on global, real-time information. An initial prototype of HUB-CI was developed and tested in the experiments. Unique features designed with the HUB-CI system include (Nair et al. 2019): (i) planned collaboration between diverse users (farmer, engineer, pathology expert, etc.) of the agricultural robotic system in a HUB-CI environment, (ii) collaborative semi-automated and manual control (remote and local) of agricultural robot, (iii) learning-based filtering algorithm for spectral images taken off plants, (iv) collaborative decision making regarding the greenhouse system based on intelligent information sharing, (v) scheduling and task administration of all cyber and human agents in the agricultural robotic system (ARS) and (vi) adaptive search and routing algorithms: use resource (time) to perform monitoring and inspection tasks. Three experiments were conducted to examine the collaborative control of the system. In all experiments, the robot was controlled from Purdue University. Two-way collaboration frames were developed: (1) an ad-hoc connection using TeamViewer in which researchers at Purdue controlled the robot's computer directly and (2) through a server using dropbox. In all experiments, collaboration with direct commands from Purdue to ARO was tested.

The hyperspectral imaging analysis can be divided into two research steps. First, a classification algorithm needs to be developed based on full spectral information of healthy and diseased spots. Second, some key hyperspectral bands need to be selected specifically for real-time in-field detection. The decrease in number of spectral bands should not affect classification accuracy. The University of Maryland research group developed a new method of hyperspectral analysis named 'outlier removal auxiliary classifier generative adversarial nets (OR-AC-GAN)' (Wang et al. 2019). The model uses full spectral information (395–1005 nm) to integrate the tasks of background removal, pixel-level spectral analysis and image-level plant classification. The model starts from generative adversarial nets (GAN) to learning the data distribution of different spectral classes. It can augment the training dataset online according to the data distribution and effectively remove the side effects of data outliers and imbalance on the dataset. This model can classify the one-dimensional spectral signal into different classes. Images were taken at ARO using a Specim hyperspectral camera with a high-resolution, high-speed image acquisition device (NI PCIe-1427) installed on an i7-4770 CPU PC. The computer was equipped with the Specim data recording application for hyperspectral images (HSI): Lumo Scanner. In the experiment for 54 independent test images of the TSWV disease database constructed by ARO, the model can reach 96.25% prediction accuracy (92.59% sensitivity, 100% specificity) before visible symptoms appear (as early as 5 days after disease inoculation) (Wang et al. 2019). In contrast, human experts can tell the difference of diseased and healthy plants 15 days after disease inoculation. For pixel-level classification accuracy, the prediction of false positives in healthy plants was as small as 1.47%. The OR-AC-GAN is an all-in-one model meeting the first requirement of hyperspectral data analysis. The experiment proved that the augmented data, a 'by-product' of OR-AC-GAN can markedly improve the performance of existing band selection algorithms (Wang et al. 2019).

8.7   Bio Inspired Robots for ARS in Precision Agriculture

Recent research has created bio-inspired robots for various agricultural applications. The fundamental motivation behind the development of bio-inspired multi-robot teams is that living organisms can successfully cope and provide good solutions to almost all robot-related problems (Tsourveloudis 2014). Navigation, material handling, sensors and machine learning are only some of the research areas that have benefited from examining and adopting methods, techniques or mimicking behaviours proved sustainable and successful for animals and humans (Tsourveloudis 2014). This section describes several bio-inspired robots that have been built for agriculturally related tasks.

Climbot (Guan et al. 2016,) is a biomimetic biped-climbing robot for potential applications in agriculture (like climbing and grasping), forestry and the building industry. Built with a modular approach, the robot consists of five joint modules connected in series and two special grippers mounted at the ends, with the scalability of changing degrees-of-freedom (DoFs). With this configuration, Climbot not only has superior mobility on multiple climbing media such as poles and trusses, but can also grasp and manipulate objects. It was inspired by observing the climbing patterns of animals such as caterpillars, chimpanzees, monkeys and sloths. Climbot may climb in several modes. The study proposed three basic climbing gaits, which are the inchworm gait, the swinging-around gait and the flipping-over gait. Autonomous climbing will be highly relevant for augmenting manual work in unstructured environments.

Guanjun et al. (2017) proposed a bio-soft robot inspired by the elephant trunk and octopus which has applications to robotic agricultural harvesting. A basic static model for axial elongation was established for the fundamental analysis of the bio-soft robot module's features, such as iso-force, isobaric and isometric characteristics.

A plant-inspired robot, named Plantoid, with sensorized robotic roots was developed by Sadeghi et al. (2016). It is the first robot prototype inspired by plants and, in particular, by the movements, sensing capabilities and behaviour of their roots. Plantoid, integrates artificial roots able to respond to environmental conditions and stimuli, performing bending movements and obstacle avoidance response. Each robotic root integrates three soft spring-based actuators that imitate the different bending capability of plant roots through variable elongation of the actuators, obtained by the direct assembly of helical springs on the shafts of DC gear-motors. Each robotic root apex embeds a matrix of commercial gravity and temperature sensors and innovative sensors for touch and humidity, customized for the specific robotic root application. The combination of sensors and a root-inspired behaviour algorithm allowed the robotic roots to move and follow external stimuli in air.

8.8 Machine Learning Applications in Agricultural CPS

An important feature of intelligence in Precision Agriculture is the ability to learn automatically from historical data and experiences (generally called 'machine learning'). Various learning methods and algorithms have been implemented in cyber physical systems, which facilitate continuous improvements, adaptations and learning from mistakes, as well as from success. Common applications of machine learning in cyber physical systems include, for example, fault detection (Sargolzaei et al. 2016), system security (Junejo and Goh 2016), pattern recognition or detection

(Spezzano and Vinci 2015), predictive maintenance (Wu et al. 2018) and adaptive scheduling (Linard and Bueno 2016).

In agricultural CPS, machine learning research (Airlanga and Liu. 2019) has addressed several Precision Agriculture topics: image classification for plant recognition, plant disease detection using hyperspectral imaging (Moghadam et al. 2017; Wang et al. 2019), smart irrigation management (Goap et al. 2018), data mining and knowledge extraction (Schuster et al. 2011; Dimitriadis, and Goumopoulos 2008), detection and prediction of biotic stresses in plants (Behmann et al. 2015; Wani and Ashtankar 2017), crop yield evaluation (Finkelstein et al. 2015, 2017), predicting environmental factors (Taki et al. 2018; Pandey et al. 2019) and automatic plant phenotyping (Yahata et al. 2017).

Future research could explore predictive maintenance, pattern detection, enhanced collaboration among agents (human or non-human agents) and system security, as related to agriculture.

8.9 Summary

Exciting capabilities and opportunities are emerging in the application of robotics in Precision Agriculture. The main areas described and illustrated in this chapter, as well as in previous chapters for robotics in different Precision Agriculture tasks include: Precision Collaboration and collaborative control (collaborative robotics), cyber physical systems, human–robot collaborative system, cloud computing, multi-robots and robot fleets, bio inspired robots, and integration of machine learning.

A summary of the dimensions of Precision Collaboration in six Precision Agriculture case studies described in this chapter is shown below (Table 8.2).

Emerging trends and future developments are planned and anticipated in all the above areas. Particular advantages can be expected by cyber-augmentation for further smart automation and autonomy (autonomation), including cyber-augmented Precision Collaboration of stakeholder farmers and human–robot agents of Precision Agriculture.

A summary of the research challenges of Precision Collaboration in different Precision Agriculture tasks is given in Table 8.3.

Table 8.2 Dimensions of precision collaboration in precision agriculture case studies

DIMENSIONS: Case Study:	Sensor-based processes	Planned collaboration	Mechanism to address or over-come or prevent errors and conflicts	Dynamic re-configuration; best matching
Robotic sonar system for specific yield assessment and plant status evaluation	Yes	No	Yes	An option
Development of a robotic detection system for greenhouse pepper plant diseases	Yes	Yes	Yes	An option
Human–robot collaborative system for selective tree pruning	Yes (for locating the cutting point)	Yes	An option	
Robot for automatic melon collection	No	Yes	An option	
Simultaneous orchard and harvesting robot design	Yes	No	No	
Selective autonomous sprayer for greenhouses	Yes	An option	An option	

Table 8.3 Summary of research challenges in Precision Collaboration and Precision Agriculture

PA Target Area	CCT approach	Challenge examples
Planting; Harvesting; Packinghouses	Humans–robots teams and swarms	• Collaborative CPS for agriculture relevant missions • Laser and sensors integration
Crops and livestock Stress monitoring, and disease detection and Prevention	Algorithms and protocols for H-R; Best matching protocols	• Sensor-based solutions • Error and conflict prevention • Fault-tolerance by teaming
Precision agriculture through cloud computing; Yield/risk estimates; Strategic and life-cycle Considerations	CDSS and RT-CDSS; Demand and capacity Sharing	• Cloud, mobile communications, e-Services for collaborative control and decision support • CPS in production, growth, and delivery
Modelling, measurement, simulation and control	DHM-R tools	• Digital production • DHM-R of ag tasks • Implications to Agricultural industry, training and education

References

Airlanga G, Liu A (2019) Initial machine learning framework development of agriculture cyber physical systems. J Phys: Conf Ser 1196(1):012065–12

Ajidarma P (2017) Multi-sensor fault tolerant learning algorithm in an agricultural robotic system. MS Thesis, Purdue University

Ajidarma P, Nof SY (2021) Collaborative detection and prevention of errors and conflicts in an agricultural robotic system. Stud Inform Control 30(1):19–28

Andújar D, Ribeiro A, Fernández-Quintanilla C, Dorado J (2016) Using depth cameras to extract structural parameters to assess the growth state and yield of cauliflower crops. Comput Electron Agric 122:67–73. https://doi.org/10.1016/j.compag.2016.01.018

Barbedo A, Garcia J (2013) Digital image processing techniques for detecting, quantifying and classifying plant diseases. SpringerPlus 2:660. https://doi.org/10.1186/2193-1801-2-660

Barbosa J, Barbosa D, Rigo S, Palazzo M, Rabello S (2014) Integrating collaborative and decentralized models to support ubiquitous learning. Int J Inf Commun Technol Educ 10:77–86

Bechar A (2010) Robotics in horticulture field production. Stewart Postharvest Review 6(3):1–11

Bechar A, Edan Y (2003) Human-robot collaboration for improved target recognition of agricultural robots. Ind Robot: Int J 30(5):432–436

Bechar A, Bloch V, Finkelshtain R, Levi S, Hoffman A, Egozi H, Schmilovitch Z (2014) Visual servoing methodology for selective tree pruning by human-robot collaborative system. In: Proceedings of the EurAgEng 2014 International conference, paper no. C0287. Zurich, Switzerland

Bechar A, Nof SY, Wachs JP (2015) A review and framework of laser-based collaboration support. Ann Rev Control 39:30–45

Bechar A, Nof SY, Tao Y (2020) Final report: Development of a robotic inspection system for early identification and locating of biotic and abiotic stresses in greenhouse crops. BARD Research Project IS-4886-16 R

Behmann J, Mahlein A-K, Rumpf T, Römer C, Plümer L (2015) A review of advanced machine learning methods for the detection of biotic stress in precision crop protection. Precis Agric 16(3):239–260

Bloch V, Bechar A, Degani A (2015) Task characterization and classification for robotic manipulator optimal design in precision agriculture. In: Proceedings of the ECPA 2015, pp 313–320. Tel-Aviv, Israel

Bloch V, Bechar A, Degani A (2017) Development of an environment characterization methodology for optimal design of an agricultural robot. Ind Robot 44(1):94–103

Bloch V, Degani A, Bechar A (2018) A methodology of orchard architecture design for an optimal harvesting robot. Biosyst Eng 166:126–137

Bock CH, Poole GH, Parker PE, Gottwald TR (2010) Plant disease severity estimated visually, by digital photography and image analysis, and by hyperspectral imaging. Crit Rev Plant Sci 29:59–107

Cheein FA, Herrera D, Gimenez J, Carelli R, Torres-Torriti M, Rosell-Polo JR, Arnó J (2015) Human-robot interaction in precision agriculture: Sharing the workspace with service units. In: IEEE International conference on industrial technology (ICIT), pp 289–295

Devadasan P, Zhong H, Nof SY (2013) Collaborative intelligence in knowledge -based service planning. Expert Syst Appl 40(17):6778–6787

Dimitriadis S, Goumopoulos C (2008) Applying machine learning to extract new knowledge in precision agriculture applications. In: Proceedings of the 12th panhellenic conference on informatics, pp 100–104

Dong X, Vuran MC, Irmak S (2013) Autonomous precision agriculture through integration of wireless underground sensor networks with center pivot irrigation systems. Ad Hoc Netw 11(7):1975–1987

Duan YE (2012) Design of agriculture information integration and sharing platform based on cloud computing. In: Proceedings of IEEE International conference on cyber technology in automation, control, and intelligent systems, pp 353–358

Dusadeerungsikul PO, Nof SY (2019) A collaborative control protocol for agricultural robot routing with online adaptation. Comput Ind Eng 135:456–66

Eizicovits D, Van Tuijl B, Berman S, Edan Y (2016) Integration of perception capabilities in gripper design using graspability maps. Biosys Eng 146:98–113

Emmi L, Paredes-Madrid L, Ribiero A, Pajares G, Gonzales-de-Santos P (2013) Fleets of robots for preciaion agriculture: a simulation environment. Ind Rob 40(1):41–58

Fermont A, Benson T (2011) Estimating yield of food crops grown by smallholder farmers. International Food Policy Research Institute, Washington DC, pp 1–68. (Open Access)

Finkelstein R, Yovel Y, Kosa G, Bechar A (2015) Detection of plant and greenhouse features using sonar sensors. Proceedings of the ECPA 2015, pp 299–305. Tel-Aviv, Israel

Finkelstein R, Bechar A, Yovel Y, Kosa G (2017) Investigation and analysis of an ultrasonic sensor for specific yield assessment and greenhouse features identification. Precis Agric 18(6):916–931

Franke J, Menz G (2007) Multi-temporal wheat disease detection by multi-spectral remote sensing. Precis Agric 8(3):161–172

Franke J, Gebhardt S, Menz G, Helfrich GH (2009) Geostatistical analysis of the spatiotemporal dynamics of powdery mildew and leaf rust in wheat. Phytopathology 99:974–984

Gebbers R, Adamchuk VI (2010) Precision agriculture and food security. Science 327(5967):828–831

Gennaro SF, Albanese L, Benanchi M, Marco SD, Genesio L, Matese A (2012) An UAV-based remote sensing approach for the detection of spatial distribution and development of a grapevine trunk disease. In Procdings of the 8th International workshop on grapevine trunk diseases, pp 734–737

Glauber JW (2013) The growth of the federal crop insurance program, 1990–2011. Am J Agr Econ 95(2):482–488

Goap A, Sharma D, Shukla AK, Rama Krishna C (2018) An IoT based smart irrigation management system using Machine learning and open source technologies. Comput Electron Agric 155:41–49

Goodwin BK, Smith VH (2013) What harm is done by subsidizing crop insurance? Am J Agr Econ 95(2):489–497

Guan Y, Jiang L, Zhu H, Wu W, Zhou X, Zhang H, Zhang X (2016) Climbot: a bio-inspired modular biped climbing robot—system development, climbing gaits, and experiments. J Mech Robot 8(2):

Guanjun B, Pengfei Y, Zonggui X, Kun L, Zhiheng W, Libin Z, Qinghua Y (2017) Pneumatic bio-soft robot module: Structure, elongation and experiment. Int J Agric Biol Eng 10(2):114

Guo P, Dusadeerungsikul PO, Nof SY (2018) Agricultural cyber physical system collaboration for greenhouse stress management. Comput Electron Agric 150:439–454

Harper N, McKerrow P (2001) Recognizing plants with ultrasonic sensing for mobile robot navigation. Robot Auton Syst 34(2–3):71–82

Herlitzius T (2017) Automation and robotics-the trend towards cyber physical systems. Agriculture Business (No. 2017-01-1932). SAE Technical Paper

Hernandez JE (2014) A reference architecture for the collaborative planning modelling process in multi-tier supply chain networks: A Zachman-Based Approach. Production Planning and Control, pp 1–17

Hillnhuetter C, Mahlein AK (2008) Early detection and localisation of sugar beet diseases: new approaches. Gesunde Pflanzen 60(4):143–149

Junejo KN, Goh J (2016) Behavior-based attack detection and classification in cyber physical systems using machine learning. In: Proceedings of the 2nd ACM International workshop on cyber-physical system security, CPSS, pp 34–43

Kassim MRM, Harun AN (2017) Wireless sensor networks and cloud computing integrated architecture for agricultural environment applications. In: 2017 Eleventh international conference on sensing technology (ICST). IEEE, pp 1–5

Khaitan SK, McCalley JD (2015) Design techniques and applications of cyberphysical systems: a survey. IEEE Syst J 9(2):350–365

Lee WS, Alchanatis V, Yang C, Hirafuji M, Moshou D, Li C (2010) Sensing technologies for precision specialty crop production. Comput Electron Agric 74:2–33

Linard A, Bueno MLP (2016) Towards adaptive scheduling of maintenance for Cyber-Physical Systems. Lecture notes in computer science, Artificial intelligence and bioinformatics, vol 9952 LNCS, pp 134–150

McBratney A, Whelan B, Ancev T, Bouma J (2005) Future directions of precision agriculture. Precis Agric 6(1):7–23

Mekala MS, Viswanathan P (2017) A survey: Smart agriculture IoT with cloud computing. In: International conference on microelectronic devices, circuits and systems (ICMDCS), pp 1–7. IEEE

Mekala MS, Viswanathan P (2017) A novel technology for smart agriculture based on IoT with cloud computing. In: 2017 International Conference on I-SMAC (IoT in Social, Mobile, Analytics and Cloud) (I-SMAC). IEEE, pp 75–82

Mizrach A (2008) Ultrasonic technology for quality evaluation of fresh fruit and vegetables in pre-and postharvest processes. Postharvest Biol Technol 48(3):315–330

Mizrach A, Bechar A, Grinshpon Y, Hofman A, Egozi H, Rosenfeld L (2003) Ultrasonic classification of mealiness in apples. Trans ASAE 46(2):397–400

Moghadam P, Ward D, Goan E, Jayawardena S, Sikka P, Hernandez E (2017) Plant disease detection using hyperspectral imaging. In: 2017 International conference proceedings of digital image computing: techniques and applications (DICTA), IEEE, pp 1–8

Moghaddam M, Nof SY (2017) Best matching theory and applications. Springer ACES Book Series

Moonrinta J, Chaivivatrakul S, Dailey MN, Ekpanyapong M (2010) Fruit detection, tracking, and 3D reconstruction for crop mapping and yield estimation. In 2010 11th International conference on control automation robotics & vision (ICARCV), pp 7–10

Moshou D, Bravo C, Oberti R, West JS, Ramon H, Vougioukas S (2011) Intelligent multi-sensor system for the detection and treatment of fungal diseases in arable crops. Biosyst Eng 108(4):311–321

Mulla DJ (2013) Twenty five years of remote sensing in precision agriculture: key advances and remaining knowledge gaps. Biosyst Eng 114(4):358–371

Nair AS, Bechar A, Tao Y, Nof SY (2019) The HUB-CI model for telerobotics in greenhouse monitoring. Procedia Manuf 39:414–421

Nie J, Sun RZ, Li XH (2014) A precision agriculture architecture with cyber-physical systems design technology. Appl Mech Mater 543:1567–1570

Nof SY (1999) Robot ergonomics: optimizing robot work. Chapter 32 in Handbook of industrial robotics, 2nd edn. Wiley, New York, pp 603–644

Nof SY (2003) Design of effective e-Work: review of models, tools, and emerging challenges. Prod Plann Control 14(8):681–703

Nof SY (2007) Collaborative control theory for e-Work, e-Production, and e-Service. Ann Rev Control 31:281–292

Nof SY (ed) (2009) Springer handbook of automation. Springer Science and Business Media

Nof SY (2019) From Integration to Augmentation, from Interaction to Collaborative Control – IE/MS Frontiers for Future Work and Factories, Proceedings of APIEMS 2019, Kanazawa, Japan, December

Nof SY, Ceroni J, Jeong W, Moghaddam M (2015) Revolutionizing Collaboration through e-Work, e-Business, and e-Service, vol 2. Springer

Nuske S, Achar S, Bates T, Narasimhan S, Singh S (2011) Yield estimation in vineyards by visual grape detection. In: IEEE/RSJ International conference on intelligent robots and systems, pp 2352–2358

Nutter FWJ, Littrell RH, Brennemann TB (1990) Utilization of a multispectral radiometer to evaluate fungicide efficacy to control late leaf spot in peanut. Phytopathology 80:102–108

Oerke EC, Dehne HW (2004) Safeguarding production—losses in major crops and the role of crop protection. Crop Protect 23(4):275–285

Oerke EC, Froehling P, Steiner U (2011) Thermographic assessment of scab disease on apple leaves. Precis Agric 12(5):699–715

Pandey A, Kumar S, Tiwary P, Das SK (2019) A hybrid classifier approach to multivariate sensor data for climate smart agriculture cyber-physical systems. In: ACM International conference proceeding series: Proceedings of the 2019 International conference on distributed computing and networking, pp 337–341

Patil JK, Kumar R (2011) Advances in image processing for detection of plant diseases. J Adv Bioinf Appl Res 2(2):135–141

Pilli SK, Nallathambi B, George SJ, Diwanji V (2014) eAGROBOT- a robot for early crop disease detection using image processing. In: Proceedings of the IEEE International conference on electronics and communication systems

Pradilla JV, Palau CE (2016) Micro virtual machines (MicroVMs) for Cloud-assisted Cyber-Physical Systems (CPS). In Internet of Things, pp 125–142

Pujari JD, Yakkundimath R, Byadgi AS (2015) Image processing based detection of fungal diseases in plants. In Proceedings of the international conference on information and communication technologies. Elsevier Science, Amsterdam, The Netherlands, pp 1802–1808

Reyes Levalle R (2018) Resilience by teaming in supply chains and networks. Springer ACES Series

Sadeghi A, Alessio M, Del Dottore E, Mattoli V, Beccai L, Taccola S, Lucarotti C, Totaro M, Mazzolai B (2016) A plant-inspired robot with soft differential bending capabilities. Bioinspirat Biomimet 12(1):

Sankaran S, Mishraa A, Ehsani R, Davis C (2010) A review of advanced techniques for detecting plant diseases. Comput Electron Agric 72(1):1–13

Sargolzaei A, Crane CD, Abbaspour A, Noei S (2016) A machine learning approach for fault detection in vehicular cyber-physical systems. In: Proceedings of the 15th IEEE International conference on machine learning and applications (ICMLA), pp 636–640

Schellberg J, Hill MJ, Gerhards R, Rothmund M, Braun M (2008) Precision agriculture on grassland: Applications, perspectives and constraints. Eur J Agron 29(2–3):59–71

Schor N, Berman S, Bechar A (2015) A robotic monitoring system for diseases of pepper greenhouse. In: Proceedings of the ECPA 2015, pp 627–634. Tel-Aviv, Israel

Schor N, Bechar A, Ignat T, Dombrovsky A, Elad Y, Berman S (2016) Robotic disease detection in greenhouses: combined detection of powdery mildew and tomato spotted wilt virus. IEEE Robot Autom Lett 1(1):354–360

Schor N, Berman S, Ignat T, Dombrovsky A, Elad Y, Bechar A (2017) Development of a robotic detection system for greenhouse pepper plants diseases. Precis Agric 18(3):394–409

Schuster R, Schulter S, Poier G, Hirzer M, Birchbauer J, Roth PM, Bischof H, Winter M, Schallauer P (2011) Multi-cue learning and visualization of unusual events. In: Proceedings of IEEE International conference on computer vision workshops, pp 1933–1940

Seok H, Nof SY, Filip FG (2012) Sustainability decision support system based on collaborative control theory. Ann Rev Control 36(1):85–100

Spezzano G, Vinci A (2015) Pattern detection in cyber-physical systems. Procedia Comput Sci 52:1016–1021

Sreeram M, Nof SY (2021) Human-in-the-loop of cyber physical agricultural robotic systems. Int J Comput Comm Control 16(2)

Stafford JV (2000) Implementing precision agriculture in the 21st Century. J Agric Eng Res 76:267–275

Stajnko D, Lakota M, Hočevar M (2004) Estimation of number and diameter of apple fruits in an orchard during the growing season by thermal imaging. Comput Electron Agric 42(1):31–42

Taki M, Mehdizadeh SA, Rohani A, Rahnama M, Rahmati-Joneidabad M (2018) Applied machine learning in greenhouse simulation; new application and analysis. Inf Process Agric 5(2):253–268 (Open Access)

Tan W, Zhao C, Wu H, Wang X (2014) An innovative encryption method for agriculture intelligent information system based on cloud computing platform. JSW 9(1):1–10

Tsourveloudis N (2014) Bio-inspired robots: learning from nature. Agent and multi-agent systems: technologies and applications. Springer, Cham, pp 1–1

US National Science Foundation, Cyber-Physical Systems (CPS), https://www.nsf.gov/pubs/2010/nsf10515/nsf10515.htm

Van der Mei R, Van den Berg H, Ganchev I, Tutschku K, Leitner P, Lassila P, Wac K (2018) State of the art and research challenges in the area of autonomous control for a reliable internet of services. Autonomous Control for a Reliable Internet of Services. Springer, Cham, pp 1–22

Wachs JP, Stern HI, Burks T, Alchanatis V (2010) Low and high-level visual feature-based apple detection from multi-modal images. Precis Agric 11(6):717–735

Wang HZ, Lin GW, Wang JQ, Gao WL, Chen YF, Duan QL (2014) Management of big data in the internet of things in agriculture based on cloud computing. Appl Mech Mater 548:1438–1444

Wang Z, Gong L, Chen Q, Li Y, Liu C, Huang Y (2016) Rapid developing the simulation and control systems for a multifunctional autonomous agricultural robot with ROS. In: International conference on intelligent robotics and applications. Springer, Cham, pp 26–39

Wang D, Vinson R, Holmes M, Seibel G, Bechar A, Nof S, Tao Y (2019) Early tomato spotted wilt virus detection using hyperspectral imaging technique and outlier removal auxiliary classifier generative adversarial nets (OR-AC-GAN). Sci Rep 9(1), Article number 4377

Wani H, Ashtankar N (2017) An appropriate model predicting pest/diseases of crops using machine learning algorithms. In: Proceedings of the 4th international conference on advanced computing and communication systems (ICACCS), pp 4–8

Wetterich CB, Neves RFO, Belasque J, Marcassa LG (2016) Detection of citrus canker and Huanglongbing using fluorescence imaging spectroscopy and support vector machine technique. Appl Opt 55(2):400–407

Wu Z, Luo H, Yang Y, Lv P, Zhu X, Ji Y, Wu B (2018) K-PdM: KPI-oriented machinery deterioration estimation framework for predictive maintenance using cluster-based hidden markov model. IEEE Access 6:41676–87

Yahata S, Onishi T, Yamaguchi K, Ozawa S, Kitazono J, Ohkawa T, Yoshida T, Murakami N, Tsuji H (2017) A hybrid machine learning approach to automatic plant phenotyping for smart agriculture. In: Proceedings of the international joint conference on neural networks (IJCNN), 1787–93

Yilmaz I, Yoon SW, Seok H (2017) A framework and algorithm for fair demand and capacity sharing in collaborative networks. Int J Prod Econ 193:137–147

Zamora-Izquierdo MA, Santa J, Martínez JA, Martínez V, Skarmeta AF (2019) Smart farming IoT platform based on edge and cloud computing. Biosyst Eng 177:4–17

Zhang L, Zhong H, Nof SY (2015) Adaptive fuzzy collaborative task assignment for heterogeneous multi-robot systems. Int J Intell Syst 30(6):731–762

Zhong H (2012) HUB-based telerobotics. M.S. Thesis, School of IE, Purdue University, West Lafayette, IN, USA

Zhong H, Nof SY (2020) Dynamic lines of collaboration - disruption handling and control. Springer, ACES Series

Zhong H, Wachs JP, Nof SY (2013) HUB-CI model for collaborative telerobotics in manufacturing. IFAC Proceedings Volumes 46(7):63–68

Zhong H, Nof SY, Berman S (2015) Asynchronous cooperation requirement planning with reconfigurable end-effectors. Robot Comput-Integr Manuf 34:95–104

Zhou L, Chen N, Chen Z, Xing C (2016) ROSCC: An efficient remote sensing observation-sharing method based on cloud computing for soil moisture mapping in precision agriculture. IEEE J Sel Top Appl Earth Observ Remote Sens 9(12):5588–5598

9 783030 770358